First Course in Digital Signal Processing using DADiSP

Allen Brown and Zhang Jun

Published 2011 by abramis

www.arimapublishing.com

ISBN 978 1 84549 502 2

abramis is an imprint of arima publishing.

arima publishing
ASK House, Northgate Avenue
Bury St Edmunds, Suffolk IP32 6BB
t: (+44) 01284 700321

www.arimapublishing.com

Contents

ii

iv

How to use this book

There are many excellent books on digital signal processing on the market includingm,

- *Digital Signal Processing: International Version: Principles, Algorithms, and Applications* by John G. Proakis and Dimitris K Manolakis (Pearson Education).

- *Discrete-time Signal Processing: International Version* by Alan V. Oppenheim and Ronald W. Schafer (Pearson Education).

- *Digital Signal Processing: A Practical Approach* by Emmanuel Ifeachor and Barrie Jervis (Prentice Hall).

- *The DSP Handbook* by Andy Bateman and Iain Paterson-Stephens (Prentice Hall).

- *Understanding Digital Signal Processing* by Richard Lyons (Prentice Hall).

- *Digital Signal Processing: A Practical Guide for Engineers and Scientists* by Steven Smith (Newnes).

For a first time adventurer into the field of signal processing, approaching many current books is quite a daunting task. Not only is the subject difficult to grasp, but many leading books on the subject are quite sizeable. If you are new student learning about signal processing for the first time you may find several of the well known titles a little inaccessible. The primary purpose of this book is to provide you with sufficient knowledge so that when you tackle the well established books on DSP, you will have a fighting chance of making good headway with them.

The second purpose is to enable the reader to gain a good working knowledge of DADiSP which is marvellous software for exploring many aspects of digital signal processing. Its highly visual aspects and its high degree of interactivity makes it ideal for gaining an understanding of signal processing. And the good news, there is a free student version of DADiSP available.

Ideally, as you read through this book, you should have DADiSP running on your PC or laptop so that you are able to replicate the simulations and displays for yourself.

Instead of just reading about signal processing you can actually perform simulations yourself and this will have the beneficial effect of enhancing your learning and understanding of the subject.

At the end of each chapter you will find a list of topics of which you should have gained reasonable knowledge. You will also find a list of DADiSP commands that you've used in the chapter.

As you progress through the book, you are encouraged to produce a print (paper copy) of your worksheets. There are a number of reasons for this practice.

- In the first instance, it gives you a permanent record of your work which should serve as a reminder of the material you have learnt. To reach each stage in the book, you will have invested time and effort and a print of your results is a record of your investment and achievement.

- Secondly, it's important you gain a good understanding of the material you work though, printing the worksheets will help your understanding.

- Thirdly in the future as you make further progress in DSP and when investigating the concepts introduced to you in this book, you will be able to refer back to the first time you encountered these concepts and you will be able to make a comparison with the levels of your new understanding.

An exciting journey lies ahead of you and as you acquire knowledge of digital signal processing you should take comfort from the fact that your level of employability will increase by the day. There is a huge demand for electronics engineers who possess a good knowledge of this subject.

Allen Brown PhD
Anglia Ruskin University, Cambridge, England

Zhang Jun PhD
Beijing Union University, China

1. Introduction to DADiSP

DADiSP is a powerful software package which can be used to great effect in learning about the many aspects of digital signal processing. A student version is available for free from,

www.dadisp.com

Before you can download the student version, you will be expected to provide details on your institution of learning. DADiSP provides a very real and visual representation of various processes involved in signal processing. As you are probably well aware by now, signal processing requires a high standard of mathematics and DADiSP is very effective at putting the maths into true context. Seeing the mathematics is one thing, being able to visualise the significance of the maths adds greatly to the learning process.

To gain the maximum benefit from this book, its recommended you read the book while in front of your PC or laptop with the student version of DADiSP running. When you encounter a command line, such as,

W2: spectrum(w1)

(which indicates the command should be written in window W2), enter it into your copy of DADiSP and execute (pressing the return key). This will confirm your own observations of the processes going on. As you progress through the book, keep using this method of comparing what you read and what you see on your own screen - your're own active participation is very important in learning about digital signal processing. You are also encouraged to make changes in the DADiSP commands in order to observe changes in the various processes to which you are introduced.

1.1 Screen Display of DADiSP

DADiSP is a highly visual software environment which has multiple windows that enable you to gain instant access to data and processes. It has a feature in common with cells in a spreadsheet, instead windows can be made to be dependent upon each other, changing one window will cause corresponding changes in the other windows Figure 1.1 shows a screen display of DADiSP with multiple windows. In the student version you are limited to eight windows whereas in the full professional version you are limited only by the size of your monitor (or monitors) - the larger the monitor the more windows you can comfortably access. You can access each window individually where you able to exercise control over almost every aspect of a window ranging from colour, trace thickness, labelling, scaling and size. You will observe at the top of the display there is a menu bar and below that a set of icons.

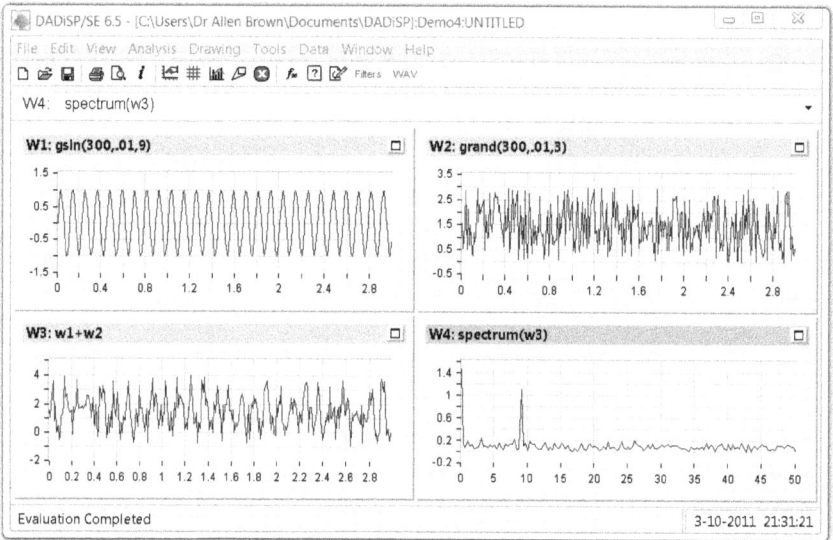

Figure 1.1: Screen display of DADiSP with
multiple windows showing individual plots which can be stand
alone or interdependent upon other windows

These can be used to navigate through the software from which you will be able to access many of the features of DADiSP. The software comes with a very useful and detailed Help File which is accessed by F1. In fact a number of features can be activated by the F Keys and these are shown below under each icon. DADiSP has scores of commands and as you progress through this book you will be introduced to many of them. After a while you should become quite skilful in using them as several of the DADiSP commands are used in many chapters in the book.

Printed Record

You also have the option of printing out an individual window or the whole display (clicking on the printer icon) and you are recommend to do this in order to keep a record of your work. Signal processing is a difficult subject and keeping a visual record of the various processes you encounter will help to consolidate your understanding of the subject.

Window Appearance

F5

The 7th icon from the left changes the appearance of the labels and scales on the plot. The default is none at all. By clicking this icon you can access the range of options. To make actual changes to the labelling you can access the Properties dialogue box which will be discussed in the next section.

Grids

F7

There are various options for adding a grid and changing its appearance. It is good practise to add a grid to a window as you may want to obtain a measurement from it.

Graph Style

F7

There are various options for displaying data plotted in a window. This icon allows you to cycle though all the options. The last option is a table containing all the data values - these can be edited as required.

Stop

Sometimes calculations may appear to be taking much longer than you would expect and it's possible your program has been caught in a loop. This icon can be used to stop the current processing.

Function Wizard

fx

There are many functions within DADiSP and one method of accessing them is this icon. This icon leads to a drop down menu showing the Analysis Functions. A particularly useful facility for easy access if you have problems in committing DADiSP function to memory.

Drawing Toolbar

Very useful if you have a need to add additional text or to draw features on a window. Clicking this icon leads to the toolbar shown below, which can be used to great effect to add text or drawing features in different colours. A pair of cross-wires are generated to allow you to position text with accuracy on a plot

Filters

Filters

This facility is used for designing digital filters and will be explored in detail in Chapters 9 and 11. Since digital filters are key devices in digital signal processing an understanding of how they are designed is crucial to a practising electronics engineer.

Audio Processing

WAV

The are numerous occasions where DSP is used in the processing of audio signals and this features allows you to import audio files into DADiSP for investigation. It will be used throughout this book when considering DSP in practical applications.

Although this is not an exhaustive discussion of all the features, it does point out the icons which you will find very useful as you become proficient in the use of DADiSP.

You may have noticed on the DADiSP website a number of tutorials. You are recommended to work through the first entitled DADiSP Product Demo as this gives an excellent introduction to the software. Later as you progress through this book you will be encouraged to work through the other tutorials as the relevant material is discussed. There are three documents which will be useful to download and they are located at,

www.dadisp.com/webhelp/dsphelp.htm

These comprise the,

- *User Guide*
- *Developers Guide*
- *Worksheet Functions.*

In the first instance the User Guide is a must and you will find the Worksheet Functions also useful. Once you become very proficient in the use of DADiSP you will find the Developers Guide quite invaluable in learning how to fashion DADiSP to your particular needs. There are many DADiSP commands and instructions and you are encouraged to discover how they are used and what benefit can be derived from their use.

1.2 Properties

When you right click on a window, a drop down menu appears and you will find more useful features which can be accessed with ease. The last entry in this menu is the Properties. This dialogue box is shown in Figure 1.2*a*. It is important to gain a good understanding of the several features in this display. This will enable you to produce traces with the colours of your choice and select appropriate labelling. Although not all the features are discussed, you will gain a good understanding of what can be achieved very quickly. As you produce traces you will probably refer to the Properties dialogue box quite often. It is therefore worthwhile spending some time learning what the software has to offer to customise your displays,

Axis

On the top left of Figure 1.2*a* you will see information relating to the **X Scale** and top right to the **Y Scale** of the plot - whether you want linear or log scaling. The scales on the plot will change automatically. You can adjust these by entering appropriate numbers in the fields. You can fix your settings by removing the tick from the fields. The choice of positioning the label is selected from the **Scales Location**: which is a drop down menu. There are several options to choose from in this menu. Spend some time viewing these options and what they have to offer. Once you've made a selection, tick the **Preserve Axis Setting** box. Without a tick in this box the scales and labels will sometimes return to the default setting. The **Autoscale** is useful if you have made changes to a window and the plot has gone off your designated scale.

Figure 1.2a: The Axis properties box

Figure 1.2b: Control can be exercised over the appearance of the grids

Grids

This range of options allow you to change the appearance of the grids on the plot as seen in Figure 1.2*b*. Keep the grid colour as grey otherwise it becomes too intrusive on the plot. You can also access a number of grid settings by clicking on the grid icon. Clicking on this icon several times allows you to cycle through the various grid setting. This is particularly useful if you have several traces in the same window and wish to change the appearance of each trace in a different style.

Data Setting

This dialogue box is shown in Figure 1.2c. You can have several traces on the same plot hence the option of selecting the Data Series. If you require a preselected unit, the drop down menus offer a wide range of options. You can also select the intervals in the x data and offset the data if required.

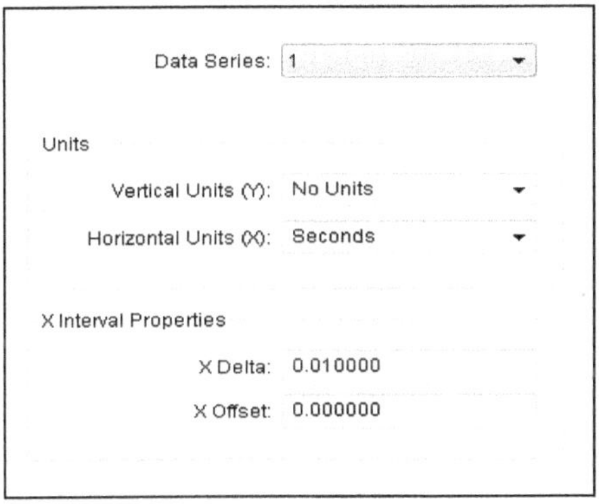

Figure 1.2c: The Data Settings options

Labels

The dialogue box for the labels is shown in Figure 1.2d.

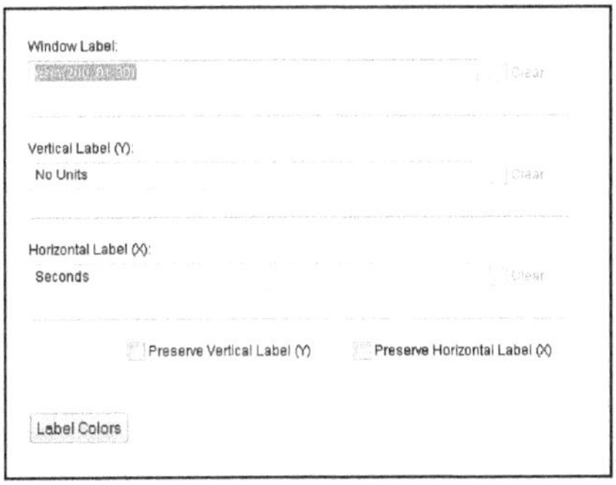

Figure 1.2d: Part of the Labels dialogue box

You can change the window label and also the Vertical Label and Horizontal Label. You will notice can also select the intervals in the *x* data and an offset if the Horizontal Label default is Seconds. To keep these labels tick the boxes otherwise they will change to their default values. You also have a wide selection of Label Colours from which to choose.

Graphical Settings

This dialogue box (shown in Figure 1.2*e*) is very useful if you have several traces overlaid in the same window - useful when you want to make a comparison. You will observe the default setting. The **Data Series** allows you to treat each trace in a different manner, use this feature to select your trace. The options in the **Settings Styles** enable you to select how the data is represented, for example lines or dots. You can also choose whether you want solid or a variety of dashed lines and finally the colour and thickness of the trace.

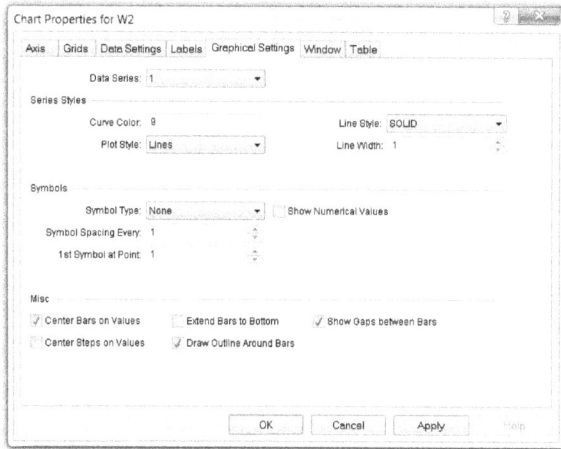

Figure 1.2e: The Graphical Settings dialogue box

A great deal of flexibility is offered the user when specifying their personal graphical settings. If you have relatively few data points you may wish to show them individually and the options in the **Symbols** section allow you control over this feature. For example in

Figure 1.2*f* you can see the data points as blue triangles with numerical amplitude values attached to each data point.

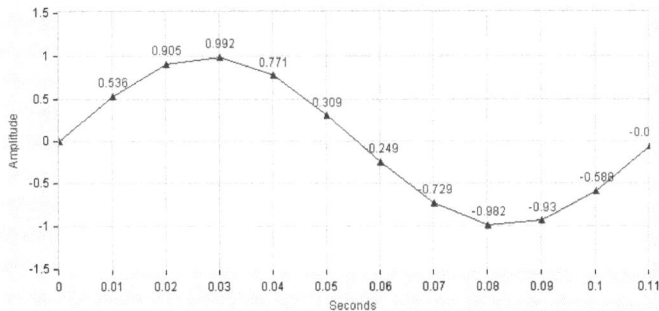

Figure 1.2f: Showing values be attached to data points

As you can appreciate this is a very useful feature if you have a plot with relatively few data points as it provides a graphical view and a data value display.

Window

This feature allows you make changes to the colours of a window including default grid intensity. Also the aspect ratio of a plot can be changed to square if required - default setting is automatic.

Table

Each window which is created in DADiSP not has the visual trace of the data but also a tabular form of the data - which is a useful feature if the data is required in a spreadsheet format or producing a suitable print out. In fact data can be imported from a traditional spreadsheet.

1.3 Data Presentation

Plots and traces are effective methods of presenting data in a variety of visual formats. The human mind likes images, this is hardly surprising because the brain has evolved several features for processing visual fields. There are several commands in DADiSP which enable a variety of visual aspects to be realised, not only 2D but also 3D plots. The customary 2D plots which appear throughout the book will be explored in some detail as you progress through the text, although most of the traces will be amplitude or magnitude vs time or frequency. With two or more traces the XY feature allows them to be plotted against each other as shown in Figure 1.3. The feature in Figure 1.3 is known as a *Lissajous Figure* where in this example one frequency is four times greater than the other. Next time you use a oscilloscope look for the feature for showing Lissajous figures, usually the XY setting. The XY option is useful when comparing the signals passing into a system and emerging from a system such as a digital filter as shown in Figure 1.3.

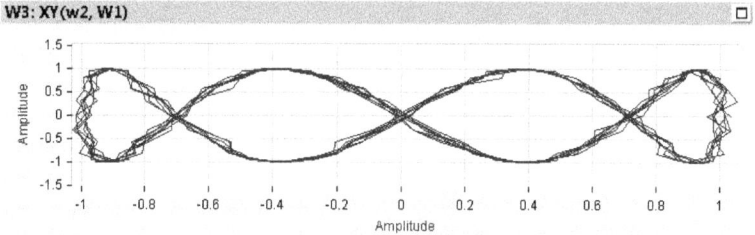

Figure 1.3: An example of an XY plot

From the Lissajous figure you can determine the phase difference between the input and output waveforms of a system for a given frequency. If the need arises to construct 3D plots DADiSP has a number of features for creating them. For example a *3D sinc function* (which is discussed in detail in Chapter 6) is shown in Figure 1.5.

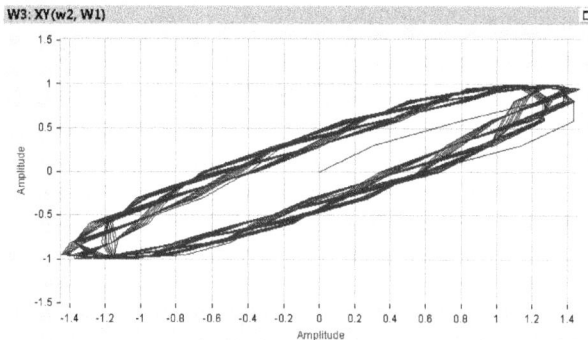

Figure 1.4: Comparing the complex input and output of a system

W1: r=sqrt(x*x+y*y); sinc(10*r);setplottype(4);setplotstyle(0)

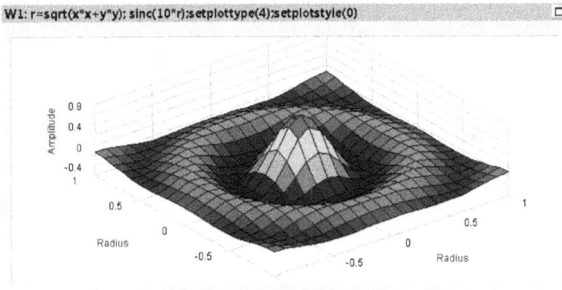

Figure 1.5: A 3D sinc function

W1: (sin(8*r)/sin(r))*cos(5.2*r);setplottype(4);setplotstyle(0)

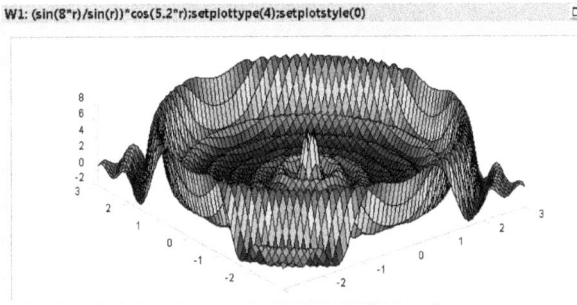

Figure 1.6: Variations of a modified sombrero

W2: plot3d(w1); mouserotate

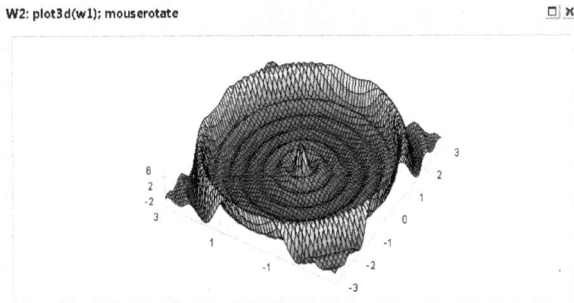

Figure 1.7: An example of the effect of the mouse rotation

The figure shown in Figure 1.5 is sometimes known as a *sombrero* - after the hat made famous in Mexico. To obtain a rotation of a 3D figure the command ;mouserotate should be included in the command line. The orientation of the figure can then be achieved with ease by mouse click and drag. An illustration of 3D figure rotation is shown in Figure 1.6. You will also observe from Figures 1.6 and 1.7 the two different colour schemes which are selected from the Properties dialogue box where there are several from which to chose. Should you be interested in generating different figures, you can start by clicking on the Function Wizard → Generate Data → Z=F(X,Y). This opens the dialogue box as shown in Figure 1.7. Select the values as shown in the dialogue box and you should arrive at an interesting 3D shape.

Figure 1.7: The dialogue box for Z = F(X,Y)

Change the values in the Z=F(X,Y) field and you should create some curious shapes - DADiSP can be fun as well! Although very interesting to look at, you may well ask do they have any functional value? The answer is yes, particularly if you perform *image processing* operations. Although not within the scope of this book you will find various instruction in DADiSP for manipulating and processing images. One is *spatial filtering* which is accomplished with a figure as shown in Figure 1.5. Having been introduced to a number of the display features in DADiSP, you are encouraged to work through the *DADiSP Workshop User Guide* which can be accessed via the Help option in the menu bar.

1.4 Concluding Remarks for Chapter 1

As in many avenues in life, the more you practise using DADiSP the greater your level of proficiency will become. You are therefore strongly encouraged to replicate the DADiSP commands so you are able to see the results for yourself on your own PC or Laptop.

Making Use of a Notebook
It's also a good idea to have a note book and keep a note of the DADiSP commands you use. Do not make the mistake of thinking you can remember them all - your memory is not as good as you think it is. One of the signs of an accomplished and successful student is their ability to know when to make notes. When learning about DADiSP you should be making notes relating to your newly acquired knowledge - this is a mark of your progress. Also, each entry in your notebook should be dated. This way you can monitor your progress. Keep a list of the commands and functions you use so you are able to gain a good understanding of how they work and the information they provide for you.

Personalising this Book
This is your copy of the book and as you progress through it, do not hesitate to use a *highlighter pen* to mark material which you deem important. This will help you when you refer back to the material in the future. The learning process is about repetition and reference to what you have already learnt. Learning about DSP is a continual process which you should find an enjoyable experience.

2. Nature of Signals

In man's understanding of nature, it is the usual custom to try and classify observations or objects into categories. Experts in Natural History have been very successful is this respect by classifying trees, plants, insects, mammals, fish - you name it - it belongs to a man made group. Signals are no different, their origin and nature enable them to be classified and this will determine how they can be processed and measured. Very often the distinction is not always clear cut and a observed signal may be a mixture of well known signals. For example a periodic signal, whose features can be accurately determined, maybe contaminated with noise which is more difficult to characterise. The most straight forward signals to characterise are deterministic signals. The most well known of these is the *sine wave* as shown in Figure 2.1 which has a characteristic wave like appearance - a common feature of many signals. To create this, open up six windows in DADiSP, and in window 1 (W1) write,

W1: gsin(200,.01,5)

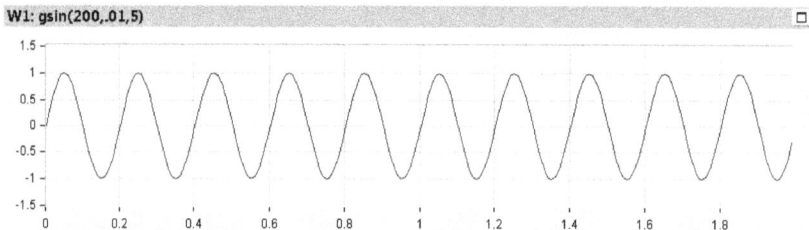

Figure 2.1 the sine wave of frequency 5 Hz

You will observe it's cyclic nature (the repetitive characteristic) and there are few parameter definitions relating to this waveform.

Frequency f: the number of repeatable cycles which occur per unit time (seconds). The measure of frequency is Hz which is equivalent to the number of cycles which occur in every second. In the not too distant past the units of frequency were *cycles per second*. When discussing rotation in machines the unit of frequency is *revolutions per minute (revs/min)*. Hz is named after the German physicist *Heinrich Rudolf Herts* who performed exciting research into the nature of electromagnetic theory.

Period T: this is the reciprocal of the frequency and is measured in seconds,

$$T = \frac{1}{f}, \text{ also } f = \frac{1}{T}.$$

Amplitude: this is a measure of the size of the signal and its units are dependent on the origin of the signal.

Heinrich Rudolf Hertz
(1857 - 1894)

Phase difference: is a measure of the possible delay or advance in a deterministic signal. These definitions come together when giving a definition in terms of a function *y(t)* changing in time *t,*

$$y(t) = A_0 \sin(2\pi f t + \theta)$$

In this expression A_0 is the amplitude, f the frequency and θ the phase difference. You will observe in Figure 2.1 the sine wave has a frequency of 5 Hz. You can verify this yourself by counting the number of cycles in one second.

2.1 Deterministic Signals

These signals are characterised by the fact their parameters (frequency and amplitude) remain constant for all time. Many signals are formed as a consequence of the *superposition principle*, which states that when several signals are mixed together they in effect form a single signal. The blended signal still contains all the characteristics of each individual signal. To illustrate this effect we are going to consider the example shown in Figure 2.2 which has two frequency components, at 5 Hz and 10 Hz.

W1: gsin(200,.01,5)+gsin(200,.01,10)

You will note in Figure 2.2 that a grid has been added to the window to add it its clarity. At this stage there is no labelling, however this will come later.

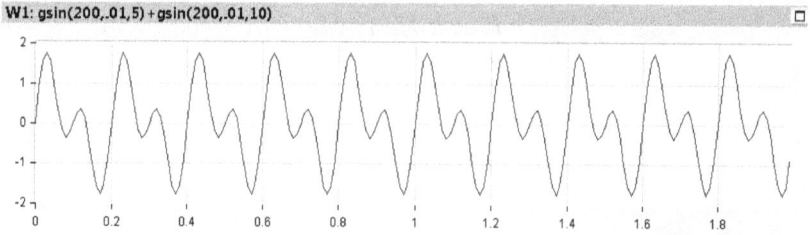

Figure 2.2: A signal with two sine waves

You could just as easily add a third frequency component at 15 Hz as shown in Figure 2.3.

W1: gsin(200,.01,5)+gsin(200,.01,10)+gsin(200,.01,15)

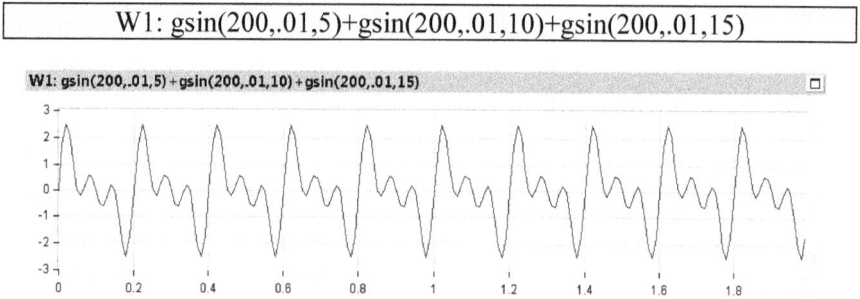

Figure 2.3: A signal with three frequency components

The three frequencies selected in Figure 2.3 are *harmonically* related. By this we mean that $f_n = nf_1$. In this expression f_1 is the *fundamental frequency*. It's a matter of convention that f_1 is also referred to as the *first harmonic*. f_2 is the *second* harmonic, f_3 the *third* harmonic and so on. The signals shown in Figures 2.2 and 2.3 are example of deterministic signals - they have fixed parameters which do not vary over time. You can use DADiSP to observe what happens to the waveform if one of the harmonics has a phase shift. Consider the following,

W1: gsin(200,.01,5)+gsin(200,.01,10,0.785)

You will observe a slight change in appearance in the two waveforms (compare Figure 2.2 with Figure 2.4).

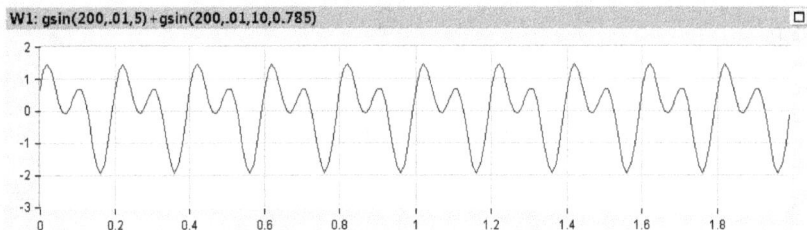

W1: gsin(200,.01,5)+gsin(200,.01,10,0.785)

Figure 2.4: Two sine waves with the second harmonics phase shifted by π/2

Exercise: Change the phase difference in W1 and observe the changes which take place in the waveform.

The waveforms so far observed are deterministic; you know what their parameters are now and also in their future. They are also called *periodic* and sometimes *stationary* owing to their periodic nature - they repeat after every cycle. Incidentally the frequency components in deterministic signals need not be harmonically related, only that the frequencies, amplitudes and phases remain constant.

2.2 Short Term Stationary Signals
Deterministic signals are generally not very interesting and rarely represent real signals encountered by the engineer. Real signals have a tendency in changing in character. An example is audio (or music). Any audio signal has varying frequencies, of changing amplitudes and phases. But what can be said about audio signals when their parameters only remain relatively constant for short periods of time. So even if you are listening to a one hundred and twenty piece orchestra playing rapidly changing notes, there are short periods when the audio signal reaching your ears is near *stationary*. These are actually called *short term stationary signals*; for audio this period over which the signal remains stationary is about 50ms.

In Figure 2.5 there is an example of an audio signal lasting just under eight seconds. You will also note there are two waveforms, one for the right channel and the other for the left channel - stereo. The effect of zooming in on a section of this signal is shown in Figure 2.6.

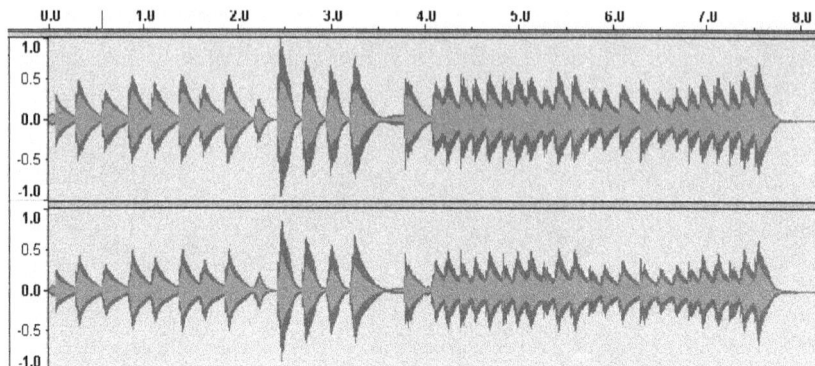

Figure 2.5: An example of an audio file

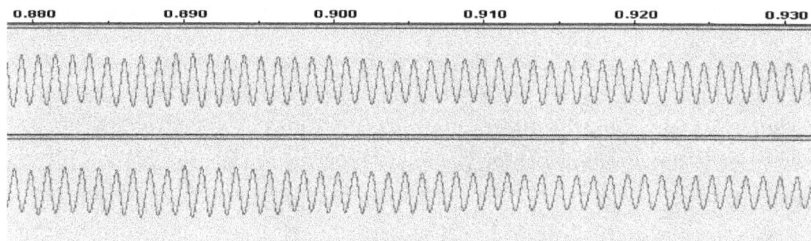

Figure 2.6: A 50 ms section of the signal shown in Figure 2.5

The duration of this sample is just over 50 ms and it can be argued the signal remains near stationary over this period. An understanding of short term stationary signals is very important when it comes to audio compression which is required when creating MP3 audio files. The assumption is for audio tracks, that for every 40ms of music there is no change in the frequency or amplitude characteristics. The analysis of the frequency content of these type of signals will be discussed in Chapter 12.

2.3 Random Signals

These are characterised by having random frequencies with random amplitudes and phases. Often referred to as noise. An understanding of random signals is vitally important in signal processing owing to the fact that most signals encountered by the engineer are contaminated by noise. In DADiSP there are various commands for generating random signals they differ by their *probability distribution function* (pdf). The pdf is a measure of how often samples appear between two defined values and is very similar to a *histogram*. As an example, first generate a random signal - 1000 sample separated by one second as shown in Figure 2.7.

W1: grand(1000,1)

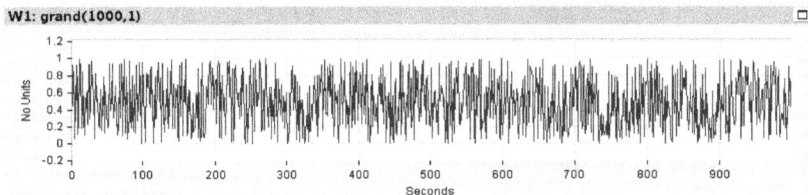

Figure 2.7: A plot of a random signal

Each value between 0 and 1 has an equal probability of occurring - a *uniform distribution*. The histogram (pdf) of the random data in Figure 2.7 is shown in Figure 2.8 distributed into 20 bins (use F7 to get the same display).

W2: histogram(w1,20)

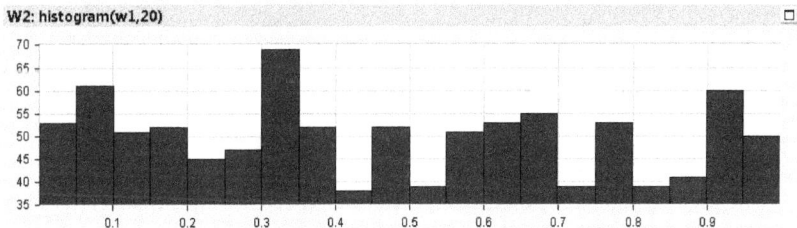

Figure 2.8: Histogram of uniform random data in Figure 2.7

Although the pdf does not appear flat (uniform), by increasing the number of random data values a uniformity will eventually appear.

Now generate a random signal of a 1,000 values with an average (mean) of 0 and a standard deviation of 0.5 as shown in Figure 2.10.

W1: gnormal(2000,1,0,.5)

W1: gnormal(2000,1,0,.5)

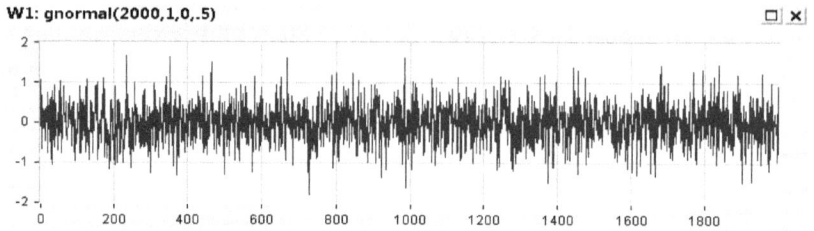

*Figure 2.9: Random data values between with a mean of 0
and a standard deviation of 0.5.*

A histogram of the data shown in Figure 2.9 can be observed in Figure 2.10 distributed into 41 bins.

W2: histogram(w3, 41)

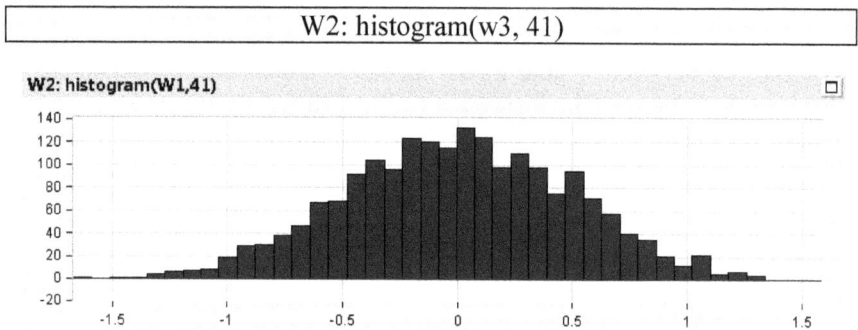

Figure 2.10: Histogram of random data shown in Figure 2.9

The profile of the data in Figure 2.10 is similar to a *Gaussian profile*. To confirm this statement, a Gaussian profile can be generated and overlaid on Figure 2.10 and this can be seen in Figure 2.11.

W3: pdfnorm(-1.5..0.01..1.5, 0, 0.5); overlay(w2)

Many of the random signals encountered by the electronics engineer have a Gaussian pdf. Incidentally, this distribution is named after the German mathematician *Carl Friedrich Gauss* who made huge contributions to the field of mathematics. The distribution function is expressed as,

$$p(x) = \frac{1}{\sqrt{2\pi\sigma^2}} e^{-\left(\frac{x-\bar{x}}{2\sigma^2}\right)^2}$$

where *p(x)* is the probability of *x* occurring when a batch of data values is sampled and σ^2 is the *variance* (a measure of the spread of the data values) of the distribution function (discussed in greater detail in the next chapter).

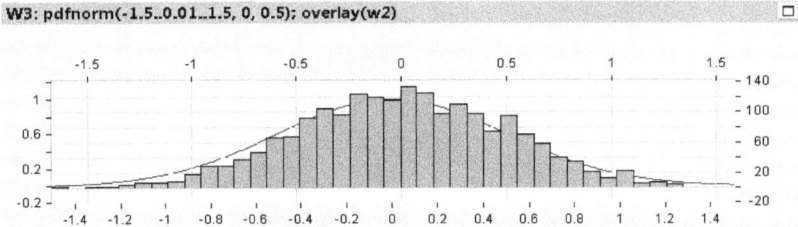

Figure 2.11: Gaussian profile overlaying the pdf of the data in Figure 2.10

This distribution function is fundamental in the field of *statistics* where random data is often encountered. When discussing terms like *average* or *mean*, this refers to the central value of a distribution which is usually the *maximum likelihood*. As an engineer you will probably come across what is known as the *central limit theorem*. This states that when batches of samples are taken from a population of samples, the averages of the batches will form a Gaussian distribution.

Carl Friedrich Gauss
(1777-1830)

2.4 Noise Contaminated Periodic Signals

As previously mentioned, many signal encountered by the engineer are periodic which are contaminated by noise (a random signal component). This occurs often when a periodic signal is passed through a noisy channel or system which imposes noise onto it. An obvious example is the propagation of a signal along a telephone cable, if there are poor connexions along the route between relay points, this will give rise to noise which gets added to the voice signal. We can simulate a noisy signal in DADiSP; start by creating a periodic signal as shown in Figure 2.13,

W1: gsin(200,0.01,4)+4*gcos(200,.01,6)

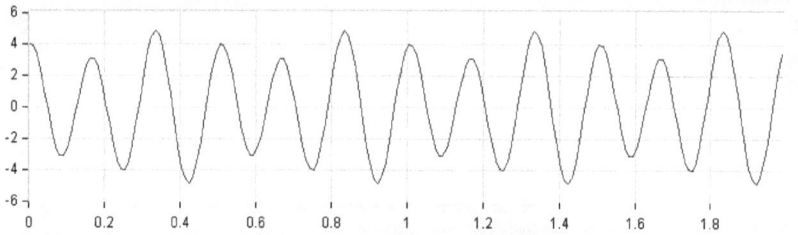

Figure 2.12: A periodic signal containing two frequencies

Now create a random signal whose *mean* is 0.6 and *standard deviation* (*std*) is 0.3 with a sample separation of 0.01 seconds as shown in Figure 2.14.

W2: gnormal(200,0.01,0.6,0.3)

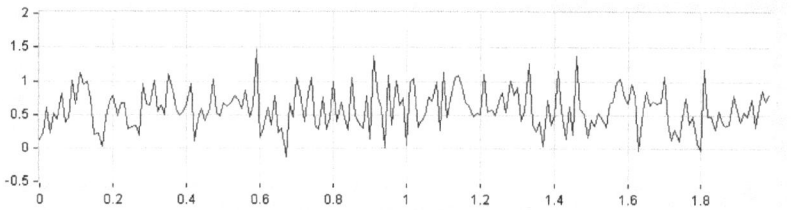

Figure 2.13: A random signal with mean 0.6 and variance 0.3

Now combine the signals with a weighting of 5 on the noise signal as shown in Figure 2.14.

W3: w1+5*w2

A comparison can be make between Figure 2.12 and 2.14. The periodicity is clearly present in Figure 2.14 together with the random component of the noise. Although the periodic aspect of the waveform in this example is discernible this is not always the case. Very often the signal is completely lost in the noise; other processes therefore have to be used to determine the periodic component.

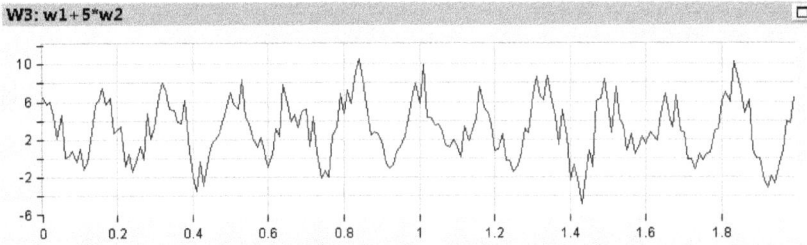

Figure 2.14: An example of a noise contaminated periodic signal

When considering the autocorrelation function in Chapter 3, extensive use will be made of these types of signals.

2.5 Transient Signals

The last class of signals to be considered are referred to as transient signals. The dictionary definition of transient is,

Transient (**tran**-zi-ent) *adj.* passing away quickly, not lasting or permanent. *n.* a temporary visitor etc.

As you can infer from this definition, a transient signal does not last very long and has a short duration. Transient signals are often encountered by engineers. For example if a bearing in a rotating machine is damaged (cracked), as the shaft rotates, every time it impacts on the cracked region of the bearing there will be a burst of acoustic signal of transient duration. As there is no function in DADiSP to generate a transient signal it can be created by using other functions. First create a periodic signal as shown in Figure 2.15.

W1: gsin(200,.01,6)+2*gcos(200,.01,12)

Now reduce the duration of this signal be applying a damping exponential function.

W2: gexp(200,.01,-4)

Once this is created, multiply W1 by W2 and the result is as shown in Figure 2.16.

23

W3: w1*w2

W1: gsin(200,.01,7)+2*gcos(200,.01,12)

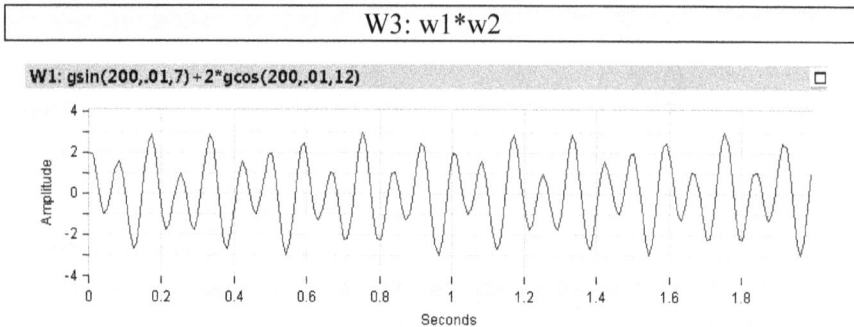

Figure 2.15: A periodic signal with two frequencies

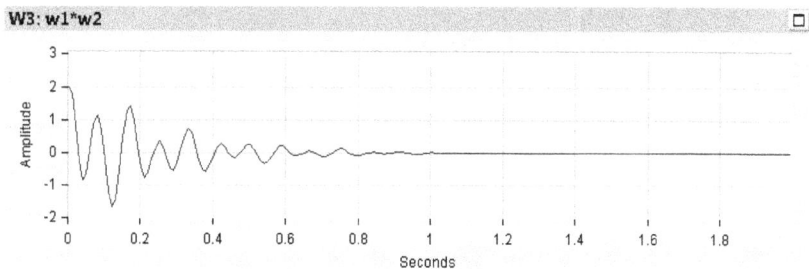

W3: w1*w2

Figure 2.16: A damped periodic signal

Figure 2.16 may be consider a transient signal as it decays in a short period. However to make the simulation more realistic transient signal, reverse the signal in W3 as shown in Figure 2.17.

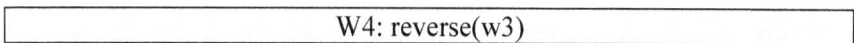

W4: reverse(w3)

You will notice that zero time is now located on the far right on the trace in Figure 2.17. One of the very useful commands in DADiSP is the feature which allows waveforms from different windows to be joined together - to *concatenate*. The traces in Figures 2.16 and 2.17 can be concatenated to give the signal shown in Figure 2.18. You will observe the *x*-axis (time) has zero time at the centre - what is required is a time offset.

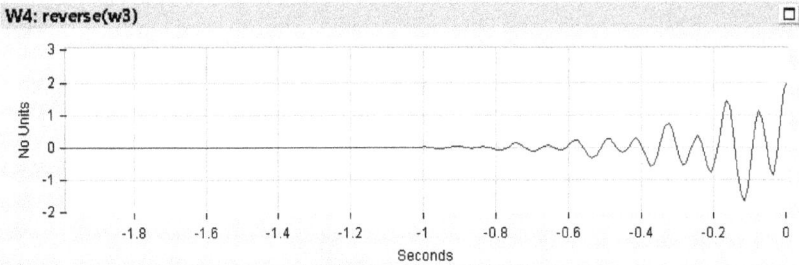

Figure 2.17: The time reversal of the signal shown in Figure 2.16

W5: concat(w4,w3)

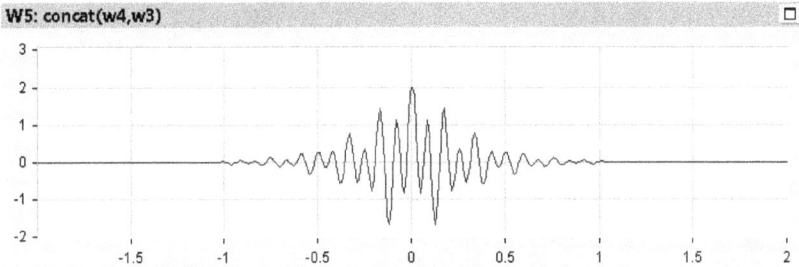

Figure 2.18: An simulated example of a transient signal

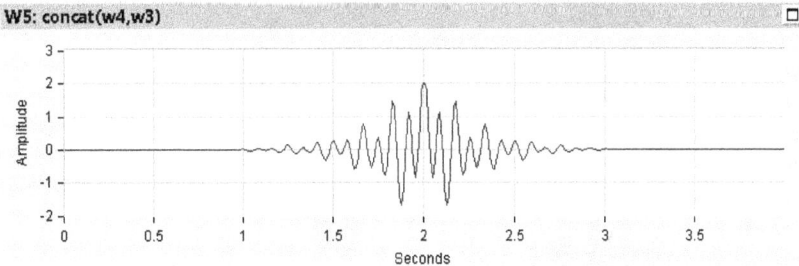

Figure 2.19: Effect of adjusting the X-Offset

The *x*-axis scale can be adjusted by navigating to the **Properties** box, select **Data Settings** and change **X Offset** from -1.99 to 0 to give Figure 2.19. Some transient signals are self windowing - they start with zero and

25

end with zero (the significance of windowing become apparent in Chapter 12). Spectral analysis can be performed directly on transient signals without any pre-processing. Another example where signals have a transient nature is radar detection of airborne platforms (aeroplanes and missiles). As the rotating antenna sweeps through a volume of sky there will be a transient reflexion of the signal which has to be processed in order to determine position, direction and velocity.

Although we have encountered several types of signals, this is by no means an exhaustive review of the different types of signals that an engineer will encounter. It does however serve as a useful introduction to the nature of signals and you will find that some signals cannot be neatly grouped into one category or another. Experience will teach you how to recognise the different types of signal you will encounter in your career.

What you have gained from this Chapter
1. Specifications of sine waves.
2. Nature of deterministic and periodic signals
3. Short term stationary signals and where to find them.
4. Creating random signals and looking at their pdfs.
5. Recognising and simulating noise contaminated signals.
6. Simulating transient signals.

The DADiSP skills you have acquired from this Chapter
1. *gsin* - creating a sine wave
2. *grand* - creating a waveform of random numbers
3. *histogram* - obtaining a histogram of random numbers
4. *gnormal* - creating a waveform of normally distributed random numbers
5. *pfdnormal* - obtaining a pdf of a random waveform
6. *overlay* - overlaying one plot over another
7. *gcos* - creating a cosine waveform
8. *W1+0.5*W2* - adding the contents of two windows
9. *gexp* - creating a exponential waveform
10. *W1*W2* - multiplying two waveforms together
11. *reverse* - reversing the layout of a waveform
12. *concat* - joining waveforms end to end

3. Tools for Processing Analogue Signals

In the previous chapter several types of signals were introduced and generally different types of processing tools used in order to derive useful information from them. It should be remembered that many digital processes have their origins in analogue processes. It's therefore important to have a good understanding of these analogue processes in order to proceed into the digital domain.

3.1 The Fourier Transform

The Fourier Transform (FT) is a well established method for deriving frequency information contained in an analogue signal. In effect the FT separates out the frequencies in a similar manner to that of a prism which refracts incident light into its constituent colours. It can be argued that a prism is a Fourier Transformer except it processes light rather than low frequency electrical signals. The FT has its origin in the *Fourier Theorem* which states that every periodic signal is a superposition of its fundamental frequency and its harmonics - each harmonic weighted differently. This was established by the French mathematician *Jean Baptiste Joseph Fourier*. The now well known Fourier's Theorem can be expressed as,

Joseph Fourier
1768 – 1830

$$x(t) = \frac{a_0}{2} + \sum_{k=1}^{k=\infty} [a_k \sin(2\pi f_k t) + b_k \cos(2\pi f_k t)] \qquad \text{... 3.1}$$

where

$$a_0 = \int x(t)dt, \ a_k = \int x(t) \sin(2\pi f_k t)dt \qquad \text{... 3.2}$$

and

$$b_n = \int x(t) \cos(2\pi f_k t)dt \qquad \text{... 3.3}$$

The coefficients $\{a_k\}$ and $\{b_k\}$ can be expressed in complex notation by introducing,

$$c_n = a_k - jb_k \qquad \text{... 3.4}$$

where $j = \sqrt{-1}$, therefore

$$c_k = \int x(t)\cos(2\pi f_k t)dt - j \int x(t)\sin(2\pi f_k t)dt \qquad \text{... 3.5}$$

This expression may be written as,

$$c_k = \int x(t)[\cos(2\pi f_k t) - j\sin(2\pi f_k t)]dt \qquad \text{... 3.6}$$

or

$$c_k = \int x(t)e^{-(2\pi f_k t)}dt \qquad \text{... 3.7}$$

where $\{c_k\}$ represents the size of the Fourier coefficients expressed in complex notation which relates to the size of the harmonics of f_I. Now consider the case when the fundamental frequency f_I tends towards zero. Then $f_k \to f$ and $c_k \to F(f)$ leaving,

$$F(f) = \int_{-\infty}^{\infty} x(t)e^{-2\pi ft}dt \qquad \text{... 3.8}$$

which is the Fourier Transform. Although this is not a rigorous derivation, it nonetheless indicates how the FT can be derived from the Fourier Series. In Chapter 12 you will find a discussion on the mechanism how the FT works as a process for dispersing frequency components contained within a signal *x(t)*. You will observe that a perfect FT can only be achieved providing the integration in Eq: 3.6 is performed over an infinite time. In practise the FT is performed by using the following,

$$F(f) = \tfrac{1}{T} \int_{0}^{T} x(t)e^{-2\pi ft}dt \qquad \text{... 3.9}$$

where T is the time over which the integration is performed - preferably for as long as possible. This might not be practicable as *x(t)* may not be *stationary* for too long a period of time for the FT to produce accurate measurements of magnitude. As an example of the FT, progress through the following DADiSP exercise. Open up DADiSP and create six windows, now create a signal with three frequency components; 4.6 Hz, 7.4Hz and 15.6 Hz as shown Figure 3.1.

W1: gsin(400,.01,4.6)+gsin(400,.01,7.4)+gsin(400,.01,15.6)

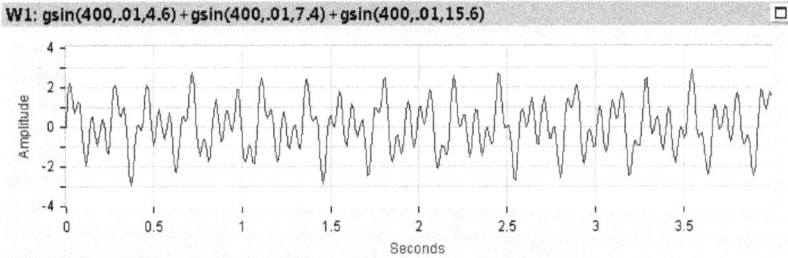

Figure 3.1: A periodic signal containing three frequencies

The spectral content of the signal can be obtained by using the *spectrum* command and the result is shown in Figure 3.2.

W2: spectrum(w1)

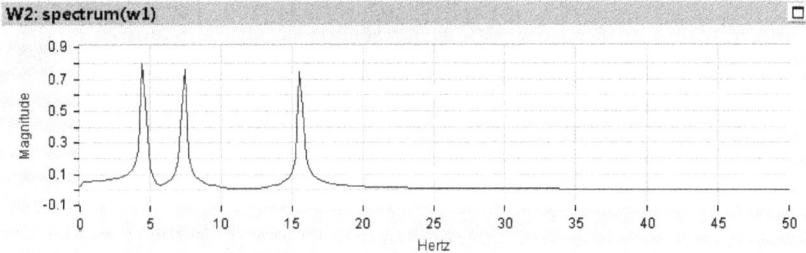

Figure 3.2: The spectrum of the signal shown in Figure 3.1

As you can observe the three spectral lines are clearly visible. One of the effects of performing the FT over a finite period is the broadening of the spectral lines and this is also evident in Figure 3.2 - the lines have a finite spectral width.

3.2 Statistical Tools for Processing Random Signals

In the previous chapter a discussion was presented on the nature of random signals and how to create them in DADiSP. You should now recognise the significance of the *probability distribution function* (pdf) and its role in characterising random signals. In fact the only way to characterise random signals is to resort to the use of statistical tools.

Mean or Average

The average or mean which was alluded to for creating the data in Figure 2.11. Given a random signal *x(t)* of duration T, the mean is expressed as,

$$\overline{X} = \tfrac{1}{T} \int_0^T x(t)dt \qquad \qquad ...\, 3.10$$

If the signal equally distributed around the zero value, the mean will therefore be very close or equal to zero. You can test this by creating a random signal and measuring the mean. The signal is shown in Figure 3.3.

W1: gnormal(500,.01,0)

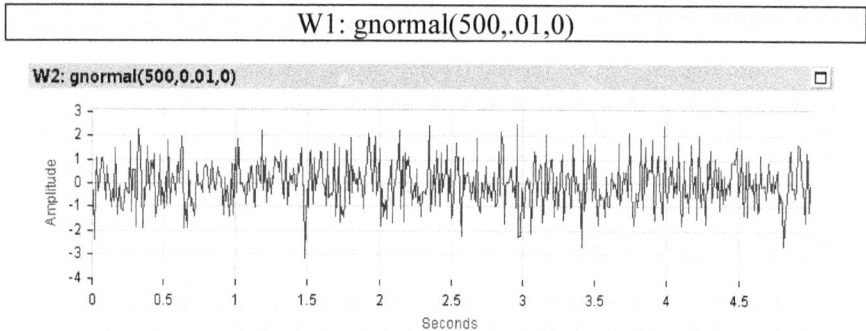

Figure 3.3: A random signal comprising 500 data values

The mean can be determined by using the mean command,

W2: mean(w1)

You will see on the bottom left of the screen a value close to 10^{-17} which is a very small number. Since in many cases the mean produces a value of zero which to the engineer is not particularly useful.

Mean Squared

A better measure is the mean squared. Consider the random signal with only 100 data values shown in Figure 3.4.

W3: gnormal(100,1,0,.5)

Now square the function by multiplying it by itself, this produces the plot as shown in Figure 3.5.

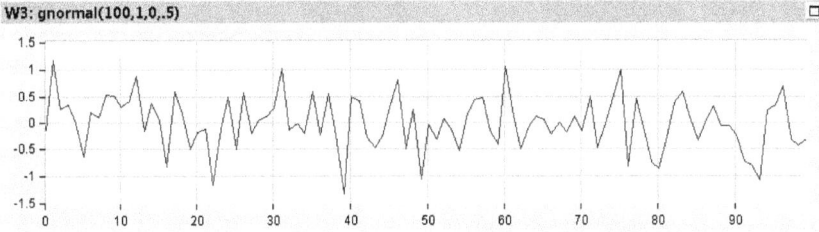

W3: gnormal(100,1,0,.5)

Figure 3.4: A random signal with only 100 data values.

W4: w3*w3

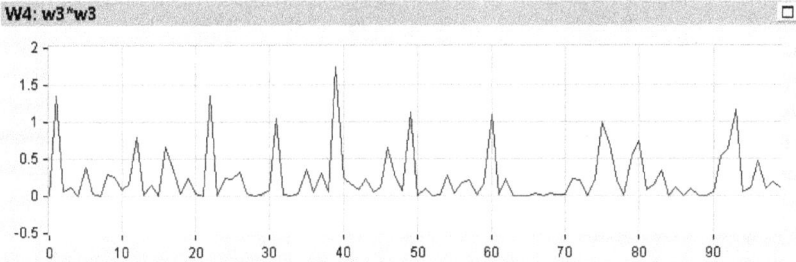

W4: w3*w3

Figure 3.5: The squared version of the signal shown in Figure 3.4

You will observe from Figure 3.5 the data values are now all positive - all above the zero axis. An integration is therefore performed on all the positive area enclosed by the curve. The mean squared is defined as,

$$\overline{X^2} = \tfrac{1}{T} \int_0^T x^2(t)dt \geq 0 \qquad \text{... 3.11}$$

It's customary to use the *root mean square* which is defined as,

$$rms = \sqrt{\overline{X^2}} \qquad \text{... 3.12}$$

When you use a *multimeter* in the laboratory and make a measure of an AC signal, it displays the *rms* value. Now perform this measurement,

W3: rms(w3)

This will give an answer close to 0.497 (bottom left of the screen). Sometime the mean of a signal is not zero, especially this true if there is a DC bias on the signal, such as the signal shown in Figure 3.6.

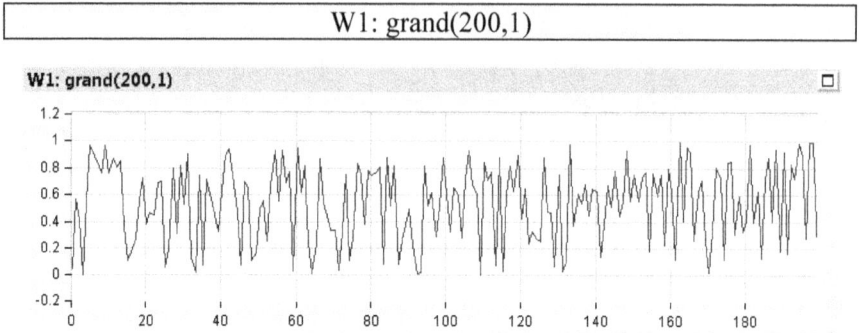

W1: grand(200,1)

Figure 3.6: A random signal whose mean is ≠ 0

The values are all greater than zero and the mean is given by,

W2: mean(w1)

which will give an answer close to 0.538.

Variance
The *variance* of a signal is defined as,

$$\sigma^2 = \overline{X^2} - \overline{X}^2 \qquad \dots 3.13$$

and the *standard deviation* (std) is given by,

$$std = \sqrt{\sigma^2} \qquad \dots 3.14$$

Summary Statistics of W1

Sample Size	200
Mean	0.456717
Variance	0.0826249
Std. Deviation	0.287446
Std. Error	0.0203255
Maximum	0.988189
Minimum	0.0104373
Range	0.977752
RMS	0.539261

OK

A measure of the std is very close to 0.289 which can be found when you click on Analysis from memu bar and select Statistics → Summary Stats... you will see a table similar to the one shown on the left which contains all the measurements just discussed. Variance is the measure of the *spread* of values in a signal. When

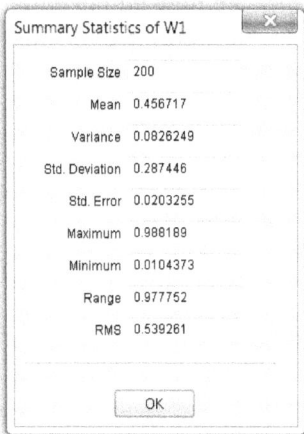

looking at a Gaussian curve (also known as a *normal distribution*), the width is often given by *full width half maximum* (FWHM) which has the approximate value,

$$FWHM = 2\sqrt{2\ln(2)} \approx 2.354\sigma$$

This can be illustrated by generating a Gaussian profile (zero mean and a variance of 1) as shown in Figure 3.7.

W1: pdfnorm(-10..0.01..10, 0, 1)

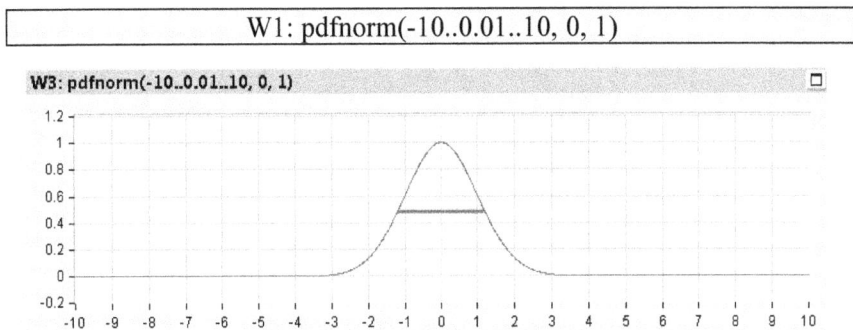

Figure 3.7: A normal distribution with a FWHM of 2.35

The width of the profile is measured at its half height which is approximately 2.35. Generating random noise in software requires the use of *pseudo random generator* algorithm which means that every random number generated is dependent on the previous random number. In essence they are not truly random, but for most applications they are quite adequate. To create truly random numbers it is possible to use a germanium diode in reverse bias connected to an *analogue to digital converter*. Germanium diodes (and germanium transistors) are very noisy and the noise mechanism within them is known as *shot noise* (random movement of electrons) which is totally random and can be used as a noise source.

3.3 Autocorrelation Function

This is a useful process for eking out the periodic component from a noise contaminated signal. Given a signal *x(t)*, the autocorrelation function *C(τ)* is defined by,

$$C(\tau) = \int_{-\infty}^{+\infty} x(t)\,x(t-\tau)dt \qquad \ldots 3.15$$

In this expression $x(t - \tau)$ represents the *history* of the signal and the autocorrelation function is able to find patterns present in $x(t)$ - patterns are periodic signals because they are repetitive. The reason why this process is successful is the fact that noise is uncorrelated and this process only finds correlated signals. Consider the signal,

$$x(t) = y(t) + n(t) \qquad \ldots 3.16$$

where $y(t)$ is the periodic signal and $n(t)$ is the noise, then

$$C(\tau) = \int_{-\infty}^{\infty} x(t)x(t - \tau)dt$$
$$= \int [y(t) + n(t)][y(t - \tau) + n(t - \tau)]dt \qquad \ldots 3.17$$

Expanding the integral will result is four separate integrals,

$$\int y(t)y(t - \tau)dt, \quad \int y(t)n(t - \tau)dt,$$
$$\int n(t)y(t - \tau)dt, \quad \int n(t)n(t - \tau)dt \qquad \ldots 3.18$$

Since the noise is not correlated with the signal, the second and third integrals are zero and since the noise is not correlated with itself the integral,

$$\int n(t)\, n(t - \tau)dt = \sigma^2|_{\tau=0} \qquad \ldots 3.19$$

is only finite when $\tau = 0$ which is equal to the variance of the noise. The autocorrelation function becomes,

$$C(\tau) = \int y(t)\, y(t - \tau) + \sigma^2|_{\tau=0} \qquad \ldots 3.20$$

The noise is therefore only present when $\tau = 0$ and this is referred to as the *zero lag*. You will observe in Eq:3.20, only the periodic component $y(t)$ remains in the integral. The autocorrelation process has therefore been able to isolate the periodic signal from the noise. Considering that many signals encountered by the electronics engineer are contaminated by noise, you would be correct to believe the autocorrelation process is therefore a powerful tool to extract a signal from noise. To demonstrate this process create a signal with a level of noise. First create a sine wave of frequency 4Hz in W1 and revisit W2 to produce a noise signal with zero mean as shown in Figure 3.8.

W2: gnormal(600,0.01,0.0,0.3)

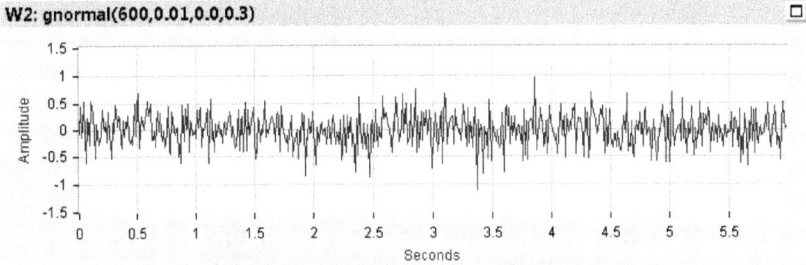

Figure 3.8: A noise signal with zero mean

Add the noise to the signal in W1 to give Figure 3.9,

W3: w1+6*w2

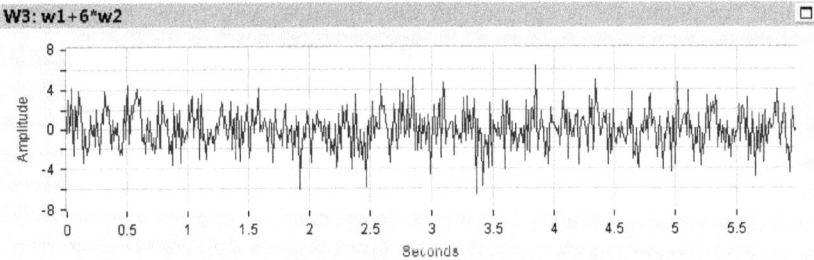

Figure 3.9: A heavily noise contaminated signal

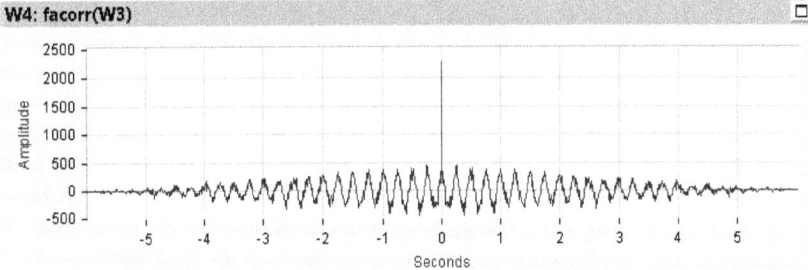

Figure 3.10: The autocorrelation function of the noise contaminated signal

There is very little difference between Figures 3.8 and 3.9 - the periodic component is well buried. Now perform the autocorrelation function as shown in Figure 3.10.

W4: facorr(w3)

You will observe from this plot the noise occurs when the *delayed time* (τ) is zero; also the periodic component is now visible. Incidentally σ² can be considered as a measure of the *noise power*. In practice the effectiveness of the autocorrelation depends on the integration time. Its definition (Eq: 3.15) requires an infinite interval for the integration to be performed which is impracticable. However when implementing the autocorrelation process, it is advisable to have as long an integration time as possible.

Effect of Finite Integration Time
Once the integration time is limited residual noise will appear in the autocorrelation function for τ ≠ 0 which is apparent in Figure 3.10. Some noise continues to be present on the signal as you move away from the zero τ value. If the noise has a non-zero mean the autocorrelation process does not work very well. You can observe this for yourself by changing W2 to have a mean > 0.

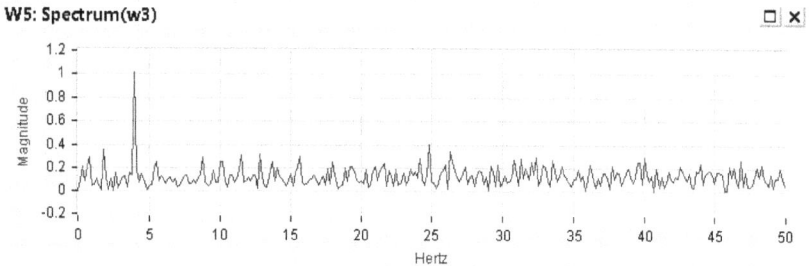

Figure 3.11: spectrum of waveform in Figure 3.9

It's instructive to see the spectra before and after the signals are correlated. The spectrum of the waveform in Figure 3.9 is shown in Figure 3.11. You will observe from this trace the periodic component is visible at a frequency of 4 Hz with noise across the whole spectral range. The spectrum of the autocorrelation function in Figure 3.10 is shown in Figure 3.12.

36

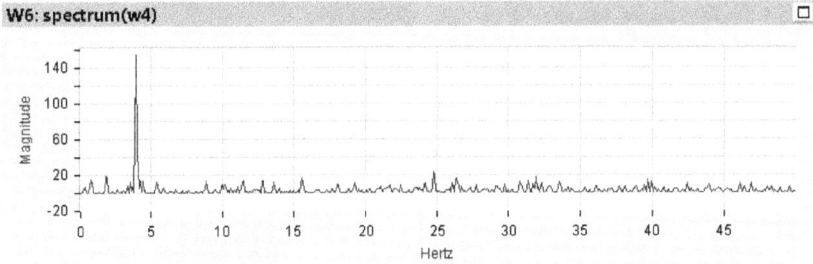

Figure 3.12: Spectrum of the autocorrelation function

You will now observe from Figure 3.12 the greater definition of the periodic component in the spectrum at 4 Hz relative to the noise which is still present across the whole spectrum but to a smaller extent. To improve on the quality of the autocorrelation function you need to increase the integration time which means, in the digital domain, a greater number of samples.

3.4 Cross-correlation

So far the autocorrelation function has been discussed - comparing a signal with its history. There is no reason why two signals from different sources cannot be compared; comparing one signal with the history of a second. This process is called *cross-correlation* and its function is defined by,

$$C_{xy}(\tau) = \int x(t)\, y(t - \tau)dt \qquad \dots 3.20$$

The signals *x(t)* ard *y(t)* are taken from separate sources and the cross-correlation function will determine whether there are any common periodic components in both signals. This is quite a common occurrence, for example a rotating machine will have many sources of vibration and the cross-correlation is used to determine whether they have common features and their physical separation. To simulate the cross-correlation process, first create a signal with three frequencies as shown in Figure 3.13.

W1: gsin(300,.01,6.1)+gcos(300,.01,2.5)-gsin(300,.01,7.6)

This generates a new signal which has three frequency components. Next create a signal in W2 which also has three frequency components, one which is also present in waveform in W1, this is shown in Figure 3.14,

W1: gsin(300,.01,6.1)+gcos(300,.01,2.5)-gsin(300,.01,7.6)

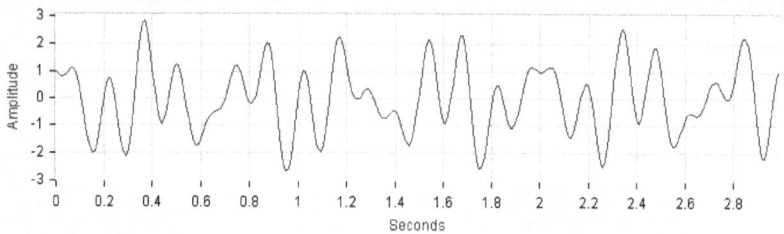

Figure 3.13: A waveform with three frequency components

W2 : gsin(300,.01,6.1)+gcos(300,.01,11)+gsin(300,.01,13.4)

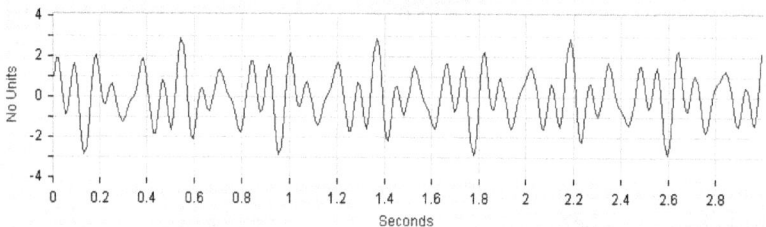

W2: gsin(300,.01,6.1)+gcos(300,.01,11)+gsin(300,.01,13.4)

Figure 3.14: A waveform with a frequency
which is also present in Figure 3.11

The cross-correlation function, located in W3, is shown in Figure 3.15,

W3 : fxcorr(w1,w2)

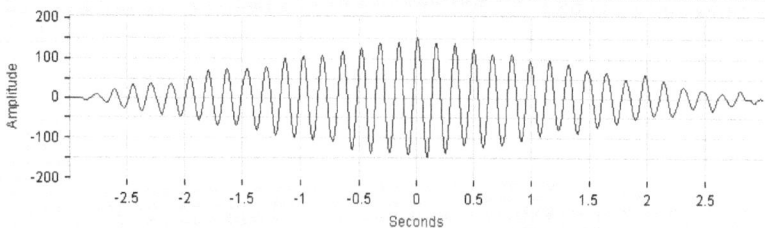

W3: fxcorr(w1,w2)

Figure 3.15: Cross-correlation of the waveform

38

The spectrum of the cross-correlation function in Figure 3.15 is shown in Figure 3.16,

W4: spectrum(w3)

You will observe the common frequency in waveforms in Figures 3.13 and 3.14 is 6.1 Hz, and as can be seen from Figure 3.16, the 6.1 Hz frequency is the only major frequency component present in the spectrum. You can verify this by tracking the mouse over W4 and right click, select Cursor → Crosshair. By dragging the mouse over the trace you will see the coordinate values (magnitude and frequency) in the top left of the screen.

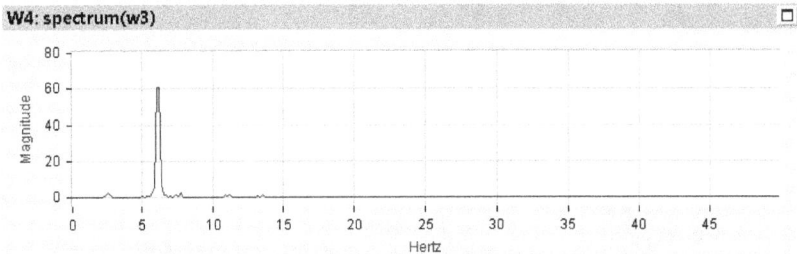

Figure 3.16: The spectrum of the cross-correlation function

The cross-correlation therefore is able to identify frequency components which are present in both *x(t)* and *y(t)*. In principle there is no reason why three or more waveforms cannot be included in a cross-correlation calculation, however the computational load can be quite heavy for such calculations. In the title of this chapter is the phrase *Processing Analogue Signals*, although these days the majority of signal processing is performed in the digital domain. Before the widespread use of digital electronic systems there were commercial systems based on analogue electronics which performed autocorrelation processes with varying degrees of success.

What you have gained from this Chapter
1. The Fourier transform and how it derived from the Fourier Series.
2. Knowledge of the statistical tools for processing random signals including mean, mean square, standard deviation and variance.
3. The autocorrelation function and how it's used to isolate a periodic signal from noise.

4. The cross-correlation function for identifying frequencies common in two waveforms.

The DADiSP skills you have acquired from this Chapter

1. *gsin* - creating waveforms with multiple frequencies
2. *spectrum* - for obtaining a spectrum of a waveform
3. *gnormal* - for generating random sample with a known std and mean
4. *mean* - calculating the mean of a waveform
5. *W3*W4* - multiplying the contents of two windows together
6. *rms* - for calculating the root mean square of a waveform
7. *std* - for calculating the standard deviation of a waveform
8. *pdfnormal* - creating a PDF of a set of random samples
9. *W1+6*W2* - adding two widows together which have differing weightings

---

DADiSP Extra

DADiSP is very effective for visualising mathematical functions. For example, $f(x) = \sin(x^3)$. Click on the Function Wizard → **Generate Data** → Y=F(X). In the dialogue box which opens, enter the following,

- Y=F(X): → $\sin(x^3)$.
- X Lower: → 0
- X Upper: → 11
- X Increment: → 0.02
- Destination: → W1.

The trace which appears in W1 is shown in Figure 3.17.

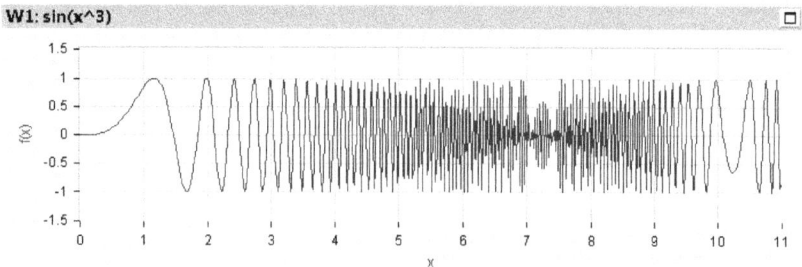

W1: sin(x^3)

Figure 3.17: A plot of the function $f(x) = \sin(x^3)$

---

4. Systems

The term system is attached to any complex configuration which comprises several components which are interconnected. To the electronic engineer a system may be electrical components designed to perform a definite task or even a set of tasks. An example is a filter which removes some frequencies and passes through others. The concept of a system is not limited to electronics engineering, it is equally applicable to other disciplines; biological, mechanical, geological, economic, political to name just a few. They have one feature in common, they are all dynamic and are subject to change. And in many cases they are subject to instability if insufficient control is not exercised on them. In this chapter the theme will be electronic linear systems and the peculiar features they possess. Generally speaking, a system is subject to inputs and it responds by changing its output(s). This can be illustrated by the diagram shown in Figure 4.1.

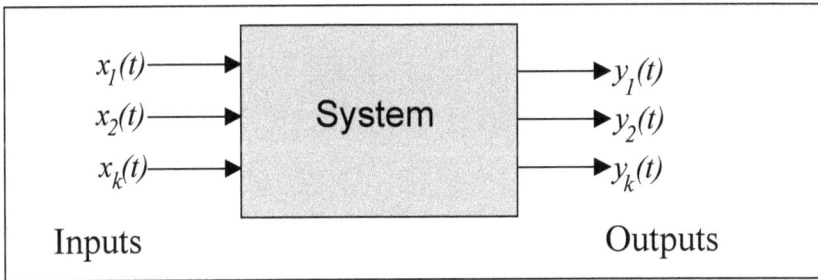

Figure 4.1: A generalised system with inputs and output responses

The system itself may comprise several subsystems depending on the complexity of the design. It is customary to characterise a system with a parameter known as the *impulse response h*. This gives a measure to how the system responds when inputs are channelled into it and its relationship to the output. One model for this relationship involves a process known as *convolution*.

4.1 Convolution

It is customary to think that a system *convolves* with the inputs to create an output. Mathematically the convolution is expressed as,

$$y(t) = \int h(t') \, x(t - t') dt' = h \star x \qquad \text{... 4.1}$$

41

In this expression *y(t)* is the output, *h* is the response of the system and *x(t)* the input. The star symbol ★ signifies *h convolving with x*. To gain an understanding of this expression, *x(t - t')* within the integral effectively represents the *current* and *all previous inputs* to the system - the system has *memory*. The current output *y(t)* is therefore made up of a weighted output from all the previous inputs to the system all added together. The weighting comes from the response function *h*. Convolution is not too different from the concept of the cross-correlation function which was discussed in Chapter 3. The convolution model is applicable to many systems in particular filters which will be considered in detail in Chapter 9. DADiSP can be used to simulate the effect of convolution. First create an input signal with three frequencies, as shown in Figure 4.2.

W1: gsin(300,.01,5)+gsin(300,.01,9)+3*gcos(300,.01,16)

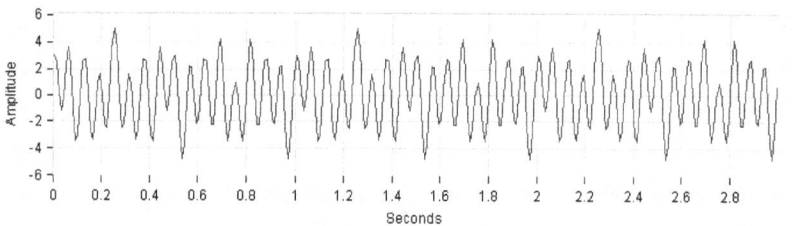

Figure 4.2: An input signal with three frequencies.

Next create a suitable response function. This can be achieved by multiplying a sine wave with an exponential damping function. The response function can be shorter, and in this case will have a duration of 0.3 seconds (30 data values) as shown in Figure 4.3.

W2: gcos(30,.01,17)*gexp(30,.01,-12)

Now perform the convolution as shown in Figure 4.4

W3: conv(w1,w2)

Figure 4.4 is in effect the output you would have from a system which has an impulse response as shown in Figure 4.3.

42

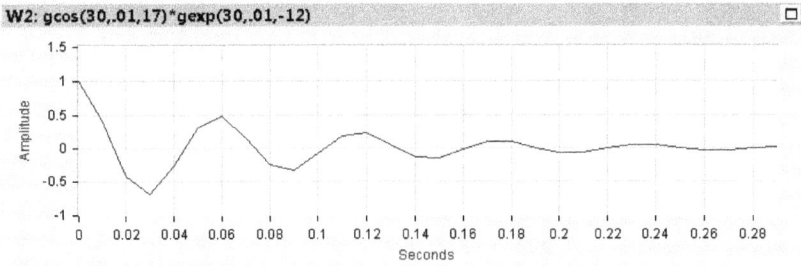

Figure 4.3: An impulse response function

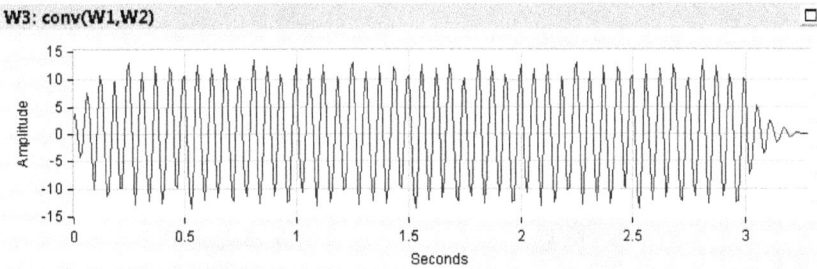

Figure 4.4: The effect of convolving Figure 4.2 with Figure 4.3

Although there does not appear to be a huge difference between Figures 4.2 and 4.3, a spectral analysis of both waveforms will be far more revealing.

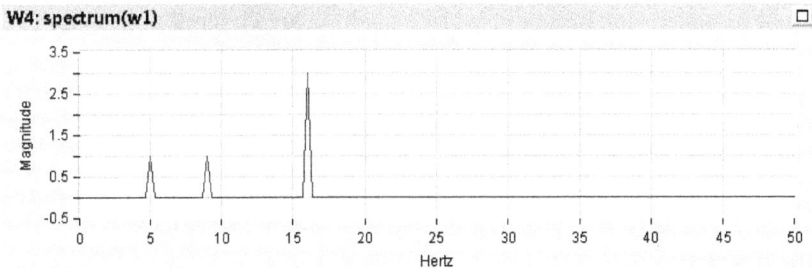

Figure 4.5: The frequency content of the signal in Figure 4.2 with the three spectral lines clearly visible.

Now obtain a spectrum of the input waveform of Figure 4.2 as shown in Figure 4.5.

W4: spectrum(w1)

Now perform spectral analysis on the output from the convolution process (Figure 4.4). This is shown in Figure 4.6.

W5: spectrum(w3)

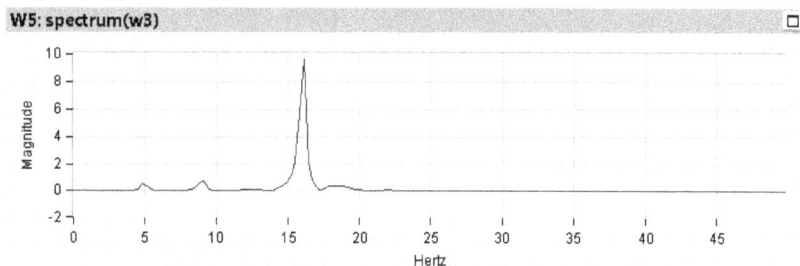

Figure 4.6: Spectrum of the output from the convolution

You will observe from Figure 4.6 there has been a serious reduction in the magnitudes of the 5 Hz and 9 Hz frequency components. Quite simply the convolution process has acted as a high pass filter by filtering out the lower frequencies and passing through the 16 Hz frequency component. From this example you are able to appreciate how the convolution process can be used as a filtering mechanism. The convolution concept is developed further to construct digital filters which are covered in Chapter 9.

4.2 Impulse Function

One of the first uses of the impulse response function was realised by the British physicist *Paul Dirac* in the early part of the last century. Dirac made significant contributions to the field of *quantum electrodynamics* and was rewarded with a Nobel Prize in 1933. The Dirac Function δ is part of that field of mathematics called *Generalised Functions* and the function was originally defined within the integral,

$$y(t') = \int_{-\infty}^{+\infty} y(t)\, \delta(t - t')\, dt \qquad \dots 4.2$$

44

This has the effect of isolating a single value $y(t')$ from the function $y(t)$ - not too dissimilar from taking a digital sample from an analogue signal. Engineers have since defined it outside the integral,

$$\delta(t-t') = \begin{array}{l} 0 \ \ for \ t \neq t' \\ 1 \ \ for \ t = t' \end{array} \qquad \text{... 4.3}$$

It is now widely used to identify the characteristics of a system and is generally called the *impulse function*. An impulse function can be created in DADiSP as shown in Figure 4.7.

W1: gimpulse(300,0.1,5)

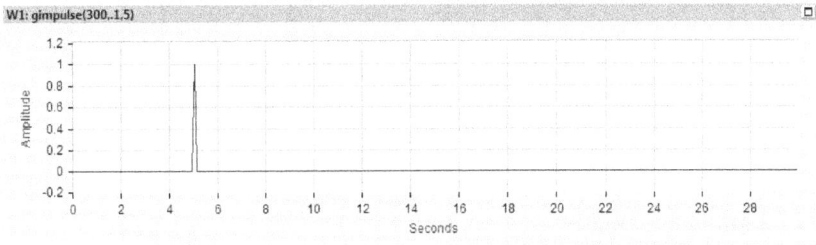

Figure 4.7: An impulse function delayed by 5 seconds

The number of data points in Figure 4.7 is 300, each separated by 0.1 second and the impulse is delayed by 5 seconds. An impulse can be characterised by a knock, or a bang whose duration is very short. Why should an impulse be of interest in signal processing? This can be answered by considering the frequency content of the function as shown in Figure 4.8.

W2: spectrum(w1)

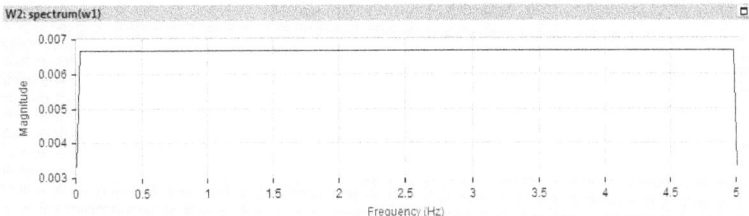

Figure 4.8: Spectrum of an impulse function

The observation which can be made from Figure 4.8 is the presence of all frequencies - it's an example of *white noise* (after sunlight which is white comprising seven colours all mixed up). Although the frequency range shown is only 5 Hz this is a feature of how DADiSP calculates the spectrum. In reality the frequency range is equal to the reciprocal of the duration of the impulse. The shorter the impulse, the greater the number of frequencies contained within it - the broader the bandwidth. By sending an impulse function, containing all frequencies, into a system the response of the system contains information relating to its spectral activity. To show the white noise content of the impulse function, perform a Fourier Transform (FT) on it. The FT is given in Eq: 3.8, substituting $\delta(t-t')$ for $x(t)$ which is the input to the FT,

$$F(f) = \int_{-\infty}^{\infty} \delta(t-t') \, e^{-2\pi jft} dt \qquad \text{... 4.4}$$

Using Eq:4.2, Eq: 4.4 becomes,

$$F(f) = e^{-2\pi jft'} \qquad \text{... 4.5}$$

The complex conjugate of this expression is,

$$F^*(f) = e^{2\pi jft'} \qquad \text{... 4.6}$$

and magnitude of the spectrum is,

$$F(f)F^*(f) = |F(f)|^2 = 1 \qquad \text{... 4.7}$$

which indicates the spectral content of the impulse function is constant for all values of f, *i.e.* white noise.

4.3 Impulse Response of a System

The main feature of an impulse function has already been mentioned in the previous section. In this section there will be more discussion on the significance of this feature. An obvious question is, how can the impulse response of a system be determined? The answer is given in it's title; it's the response of a system to an impulse input. The features of an impulse function have already been covered. If you therefore feed a system with an input comprising an impulse function, the output from the system will be its *impulse response*. This can be represented by Figure 4.9. It should be remembered the impulse function input contains all frequencies and the impulse response function output will contain all the frequencies which characterise the system. These may include resonances or spectral nulls where frequencies are filtered out. An everyday example of this process is the use of a bell. A typical bell is shown in Figure 4.10 which comprises a resonant metal body and an internal hammer.

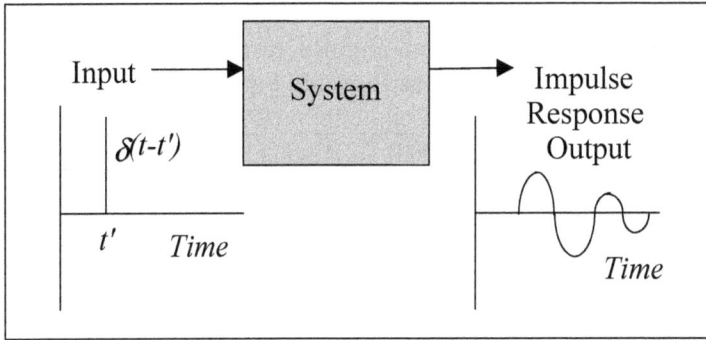

Figure 4.9: System responding to a impulse function input

The hammer, when it strikes the metal, provides the impulse function and what you hear, the resonant ringing, is the impulse response (the output) from the bell (the system). The bell will have a fundamental resonant frequency and array of other frequencies some of which are harmonically related to the fundamental. However the majority of the frequencies contained in the impulse function are filtered out as the bell does not respond to non-resonant frequencies. Most systems have similar characteristics when subjected to an impulse function and many systems have a damped ringing response function.

Figure 4.10: A bell is an example of a resonant system.

4.4 Transfer Function of a System

The impulse response of a system by itself does not provide too much information. To obtain meaningful information from the response function you need to analyse the frequency components contained within it. In this case, if $h(t)$ is the response function of a system, performing the Fourier Transform,

$$H(\omega) = \int h(t)e^{-j\omega t}dt \qquad \qquad \dots 4.8$$

$H(\omega)$ is called the *transfer function* of the system and its the complex spectrum of the system. In effect the frequency response of the system. You will have noticed that instead of using f, the radial frequency ω has been used where $\omega = 2\pi f$. To illustrate the importance of the transfer function, generate a simulated impulse response function of a system which has a resonance at 2 Hz as shown in Figure 4.11.

W1: gsin(300,0.1,2)*gexp(300,0.1,-.5)

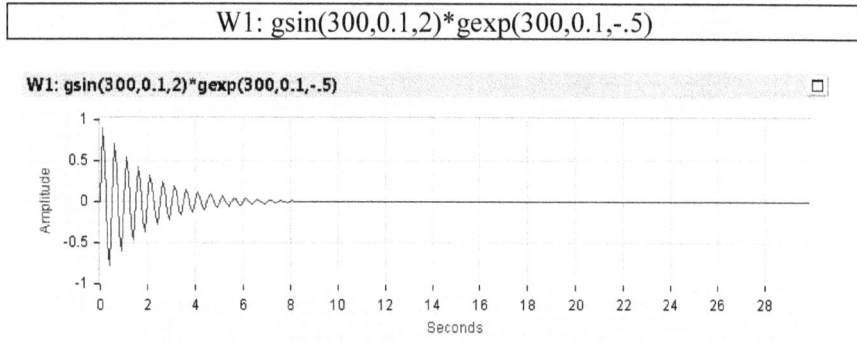

Figure 4.11: The impulse response of a system

Now determine the spectrum of this response as shown in Figure 4.12.

W2: spectrum(w1)

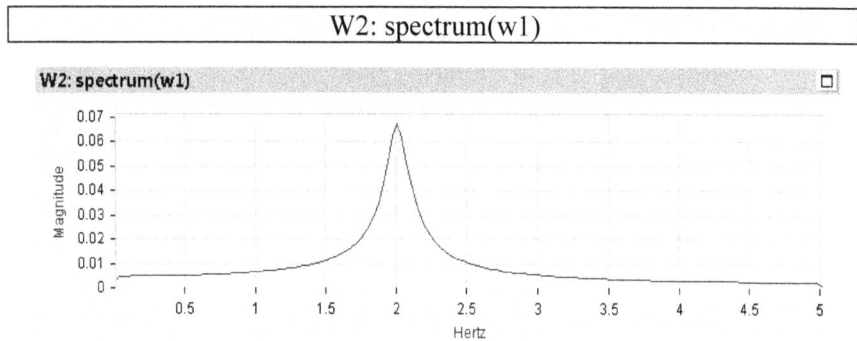

Figure 4.12: The transfer function of the impulse response as shown in Figure 4.11

You will observe what appears a broadened spectral profile of a resonant frequency at 2 Hz. Whenever a resonance is damped, it shows up in the transfer function as a broadened spectral profile. The greater the damping the broader the spectral profile. Many systems will have several resonances each with a differing degree of damping. Using the impulse function is not the only method for determining the transfer function of a system, it is quite possible to use a random noise input signal. To illustrate this process in DADiSP create a random signal as shown in Figure 4.13,

W3: gnormal(300,0.1,0,2)

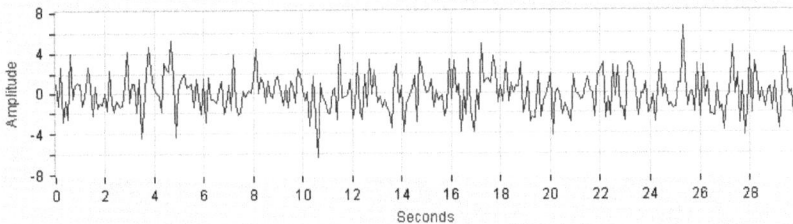

Figure 4.13: A random signal of duration 30 seconds

The data shown in Figure 4.13 has a normal distribution characteristic of noise signals encountered by the engineer. Reduce the data values in W1 to 100 which is to be used as the impulse response as shown in Figure 4.14.

W1: gsin(100,.1,2)*gexp(100,.1,-.5)

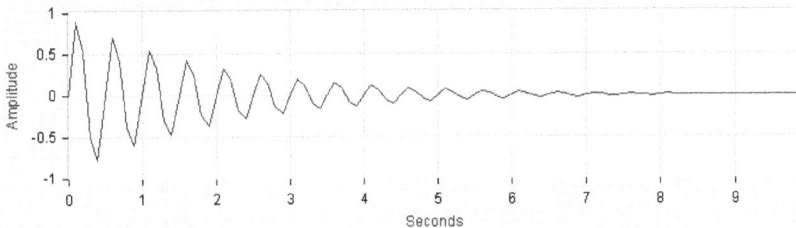

Figure 4.14: An impulse response lasting for 8 seconds

Now convolve the noise input in Figure 4.13 with the transfer in Figure 4.14 to give a impulse response function as shown in Figure 4.15.

W4: conv(w1,w3)

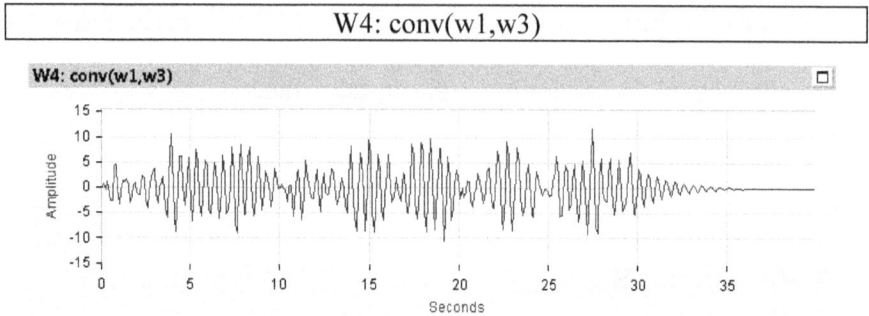

Figure 4.15: Output from the system whose impulse response is shown in Figure 4.14

The input signal has a duration of 30 seconds and you will observe the signal taking over 5 seconds to relax to zero. The spectrum of the waveform in Figure 4.15 is shown in Figure 4.16.

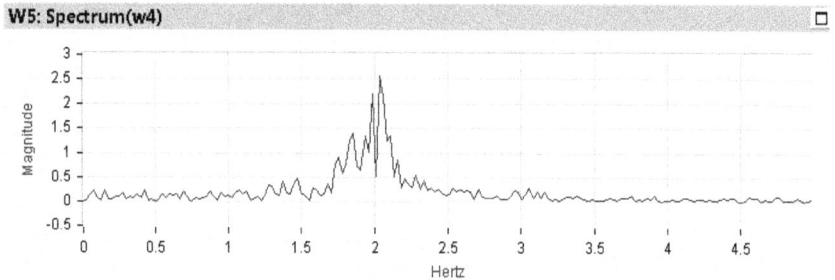

Figure 4.16 : Spectrum of the signal shown in Figure 4.15

When comparing Figure 4.12 with the Figure 4.16 you will observe the spectrum obtained from using a random signal input is not as clean as the method employing the impulse function. However, if this was to be repeated several times and then perform an average, the appearance of the spectrum would appear better defined. As an exercise, think of a method how this could be achieved in DADiSP. Very often the behaviour of a system is considered in the frequency domain. If $H(\omega)$ is the transfer function of a system and the spectrum of the input signal is $X(\omega)$, the output spectrum $Y(\omega)$ is,

$$Y(\omega) = H(\omega)\, X(\omega) \qquad \ldots 4.9$$

$Y(\omega)$ is simply the product - a straight forward multiplication, its therefore much faster to perform than convolution calculations in the time domain. In practise the type of calculation in Eq:4.9 is performed followed by the inverse Fourier Transform,

$$y(t) = \int Y(\omega)\, e^{j\omega t} d\omega \qquad \ldots 4.10$$

The calculation is performed much quicker by following this process.

Measuring Phase

If there is a need to measure the phase difference between the input $x(t)$ and $y(t)$, this task can be simulated in DADiSP. Create a sine wave of frequency 2.2Hz and convolve it with the impulse response which is shown in Figure 4.14,

W3: gsin(300,0.1,2.2)

W5: conv(w1,w3)

To determine the phase, only use part of convolved waveform which is unchanging. For this we use,

W2: extract(w5, 50. 200)

We now have two relatively stable waveforms and to obtain an XY plot, use the command,

W6: XY(w3, w2)

This produces a Lissajous figure as shown in Figure 4.17. The phase difference is related to the a and b by the following expression,

$$\theta = \tan^{-1}\left(\frac{a}{b}\right) \qquad \ldots 4.11$$

From performing a manual calculation on the plot to determine the values of a and b, we arrive at the value,

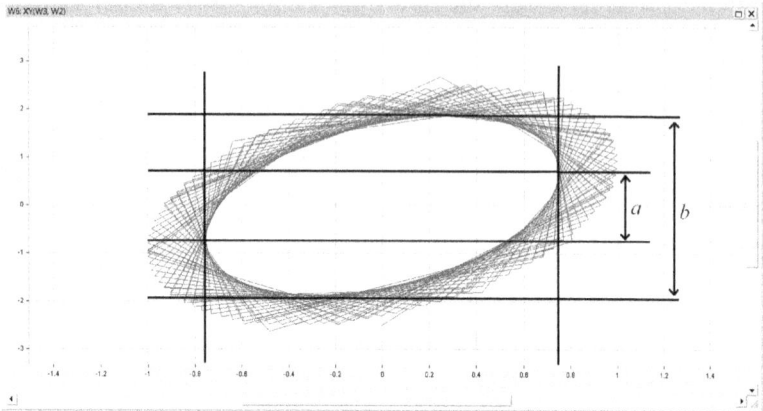

Figure 4.17: A Lissajous figure of input and output of a system

$$\theta = \tan^{-1}\left(\frac{|0.8|+|-0.8|}{|1.8|+|-1.9|}\right) = 23^{o} \qquad \text{... 4.12}$$

As the frequency is varied for the input waveform, the phase difference between the input and output waveforms will vary accordingly.

4.5 Cross-correlation of Input and Output of a System

Consider the system as shown in Figure 4.18, which has an input $x(n)$ and an output $y(n)$.

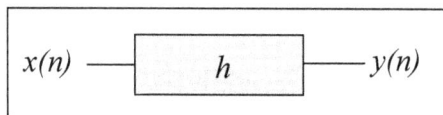

Figure 4.18: A system with input x(n) and output y(n)

At this stage we are going to introduce what is known as the *expectation value E{ }*. This is very similar to the measure of the mean. The output from the system is given by the convolution of $x(n)$ with the impulse response function,

$$y(n) = \sum_{k=0}^{k=\infty} x(k) \, h(n-k)$$

$$= \sum_{k=0}^{k=\infty} x(n-k) \, h(k) \qquad \text{... 4.13}$$

Now introduce the expectation value,

$$E\{y(n)\,x(n-v)\} = \sum_{k=0}^{\infty} E\{x(n-v)\,x(n-k)\}h(k) \qquad \text{... 4.14}$$

With no loss of generality, n can be replace by $n + v$ to give,

$$E\{y(n+v)\,x(n)\} = \sum_{k=0}^{\infty} E\{x(n)\,x(n+v-k)\}h(k) \qquad \text{... 4.15}$$

Now introduce the autocorrelation and cross-correlation functions $R_{xx}(v)$ and $R_{xy}(v)$,

$$R_{xx}(v) = E\{x(n+v)x(n\}$$
$$R_{xy}(v) = E\{x(n+v)y(n)\} \qquad \text{... 4.16}$$

Then Eq:4.15 can be expressed as,

$$R_{xy}(v) = \sum_{k=0}^{\infty} R_{xx}(v-k)h(k) = R_{xx}(v) \star h(v) \qquad \text{... 4.17}$$

By performing a Fourier Transform on this expression we get,

$$P_{xy}(\omega) = P_{xx}(\omega)\,H(\omega) \qquad \text{... 4.18}$$

which gives the transfer function as,

$$H(\omega) = \frac{P_{xy}(\omega)}{P_{xx}(\omega)} \qquad \text{... 4.19}$$

In these expressions $P_{xy}(\omega)$ and $P_{xx}(\omega)$ are referred to as the *cross power spectral density* and *power spectral density* respectively. If the input to the system is white noise, then $P_{xx}(\omega) = \sigma^2$, therefore the transfer function becomes,

$$H(\omega) = \frac{P_{xy}(\omega)}{\sigma^2} \qquad \text{... 4.20}$$

Dividing the cross power spectral density by the noise variance gives the transfer function of the system. To demonstrate this process in DADiSP, first generate a waveform comprising normally distributed values with a variance of 0.4.

W1: gnormal(300,.01,0,.4)

Use a similar impulse response function as in Figure 4.14,

> W2: gsin(100,.01,20)*gexp(100,.01,-4)

Once this response function has be created, cross-correlate with the random signal in W1.

> W3: fxcorr(w2,w1)

The result of this process is shown in Figure 4.18.

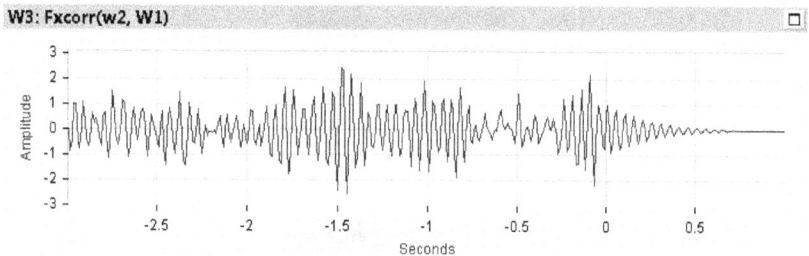

Figure 4.18: Cross-correlation of noise with impulse response function

The power spectral density of the waveform shown in Figure 4.18 is shown in Figure 4.19.

> W: psd(w3)

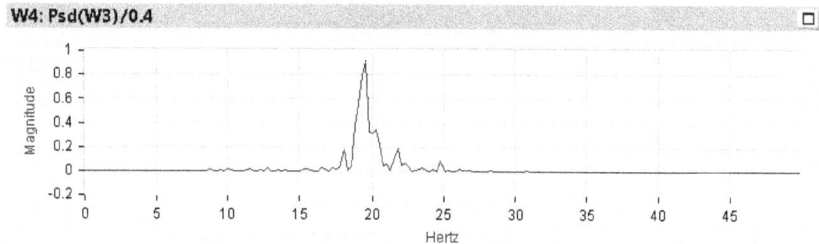

Figure 4.19: Power spectral density of the waveform shown in Figure 4.18

You will observe with Figure 4.19 the spectral line is still rather broad with a number of other features. As we are processing an input signal which is noise, its convergence towards a stable spectral value can be achieved by

---

increasing the number of samples in the input waveform. In this case, by making the following change to W1,

W1: gnormal(2000,.01,0,.4)

In effect increasing the sample number to 2000, the spectrum changes as shown in Figure 4.20.

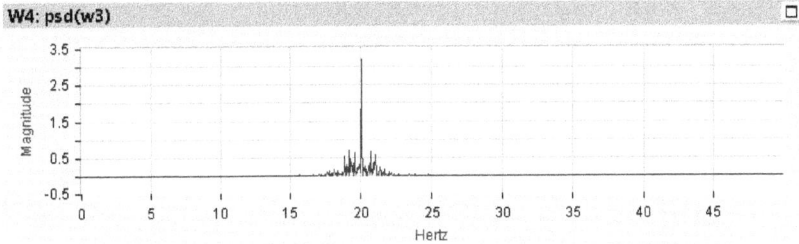

Figure 4.20: Spectral effect of increasing the number of samples

You will observe from Figure 4.20 the well defined spectral line at 20Hz which is expected. In general, the longer the duration of the input noise signal the better the estimation of $H(\omega)$. This illustrates how you can use a noise source and the ability to perform a power spectral density measurement to determine a systems' transfer function $H(\omega)$. The *Aglient 3567A* spectrum analyser shown in Chapter 14 is more than capable of performing this type of task.

What you have gained from this Chapter
1. An understanding of a system.
2. Nature of convolution
3. The significance of the impulse function.
4. An understanding of the impulse response of a system.
5. An appreciation of the transfer function of a system.
6. An insight into the cross-correlation function.
7. Using the PSD to determine the transfer function of a system.

The DADiSP skills you have acquired from this Chapter
1. *gsin* - generating a sine wave

55

2. *gcos* - generating a cosine wave
3. *gexp* - generating an exponential waveform
4. *conv* - convolving two windows together
5. *spectrum* - obtaining a spectrum of a waveform
6. *gimpulse* - generating a wavefrom with a single impulse function
7. *gnormal* - creating a waveform of normally distributed random samples with a known mean and variance
8. *fxcorr* - obtaining a cross-correlation function
9. *psd* - obtaining a power spectral density

DADiSP Extra

Continuing on the theme of visualising mathematical functions, consider the function,

$$f(x) = \sin(\tfrac{1}{x})$$

Click on the Functions Wizard → **Generate Data** → Y=F(X), when the dialogue box opens enter following,

Y=F(X): → sin(1/x)
X Lower: → -0.3
X Upper: → 0.3
X Increment: → 0.001.
Destination: → W1

You should see a trace as shown in Figure 4.21.

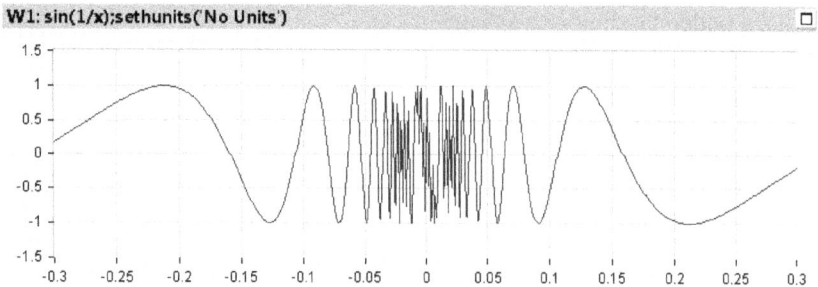

Figure 4.21: Plot of sin(1/x)

5. Digital Concepts

Signal Processing became of age when it was possible to implement DSP algorithms on digital devices, particularly Digital Signal Processors (DSPs). The processes we have discussed so far can be implemented digitally with relative ease. The majority of processes you are likely to encounter will require the use of three fundamental operations, *addition*, *multiplication* and *data storage*. DSPs are specifically designed to perform these operations at maximum speed. It is not unusual for a DSP to perform a 48-bit addition, a 24-bit multiplication and 24-bit data storage all within 10 nano-seconds (ns). DSPs have presented many opportunities to perform many signal processing tasks at high speed with low cost. The knowledge you will gain from this book will contribute greatly to your understanding of implementing the numerous processes to which you will be introduced.

Algorithm

You will encounter this word very frequently in the field of digital signal processing. You can think of an algorithm as a method for performing a process in *minimum time*. When performing processes in microelectronic systems, the execution rate is crucial - faster is better. This is where new algorithms come in - performing known or even new tasks quicker. Very often PhDs are awarded to engineering postgraduates who invent new algorithms as these represent new knowledge which is required to gain a PhD. If you are likely to pursue a career in digital signal processing you may wish to consider the significance of algorithm development.

5.1 Digitising Analogue Signals

Before a DSP can work on a signal it is necessary for the signal to be converted into a digital format. The task of converting an analogue signal into a digital signal is performed by an *analogue to digital converter* (ADC). There are several designs available ranging from the slow (dual ramp design) to the very fast (Flash Converters). The process of digitising a signal is shown in Figure 5.1 which shows an analogue signal entering a ADC and digital samples emerging in the form of binary numbers. The precision of the binary numbers which emerge from the ADC will depend on its design, ranging from 8-bit to 24-bit. The most significant bit repre-sents the sign of the binary number; 0 for a positive number or 1 for a negative number.

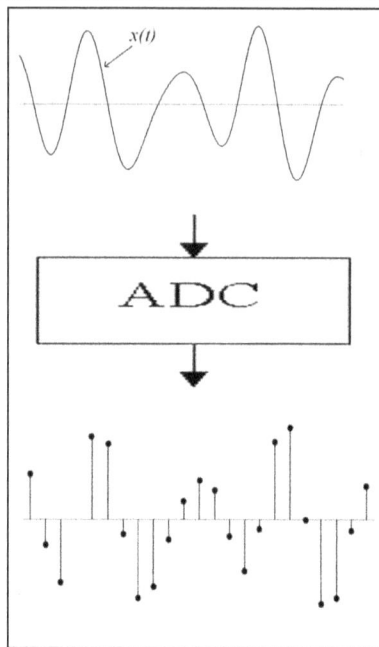

Figure 5.1: Analogue to digital conversion

Once the binary numbers are stored in memory then the digital processing begins. In many real-time applications, as each sample emerges from the ADC the digital processing begins working on the new sample. The format of the binary numbers is known as *two's complement*. The two's complement representation is used if you wish to determine the negative equivalent value of a binary number. For example, given a binary number (92_{10}),

01011100

The compliment is $(0 \rightarrow 1$ and $1 \rightarrow 0)$

10100011

Now add 1

$$\begin{array}{r} 10100011 \\ 1 \\ \hline 10100100 \end{array}$$

10100100 (-92) is the negative equivalent of 01011100 (+92). The advantage of two's complement - when added the result is zero - very useful in *Arithmetic Logic Units* (ALUs) in microprocessors which can only add numbers. When discussing ADCs there are several definitions of which you must be aware.

Sampling Frequency f_s

Measured in Hz, it's the number of digital samples produced every second. For audio applications this is 41,000 Hz for CD quality and 96,000Hz or 192,000Hz for DVD-audio.

Sampling Period Δ

This is the time interval between each sample emerging from the ADC and is the reciprocal of the sampling frequency,

$$\Delta = \tfrac{1}{f_s} \qquad \qquad ...5.1$$

Usually measured in *seconds, ms, µs* or even *ns*.

Resolution

This is a measure of the precision of the binary numbers and is the range of values which can be produced by the ADC. For an *n*-bit ADC the resolution will by 2^n.

Voltage Resolution q

The smallest voltage which can be resolved by the ADC. If the full scale input voltage is V_{fs}, then

$$q = \tfrac{V_{fs}}{2^n - 1} \qquad \qquad ...5.2$$

This is sometimes referred to as the *quantisation* level.

Aliasing

This condition can arise if the sampling frequency is too low - the input frequency to the ADC is higher than it can comfortably sample.

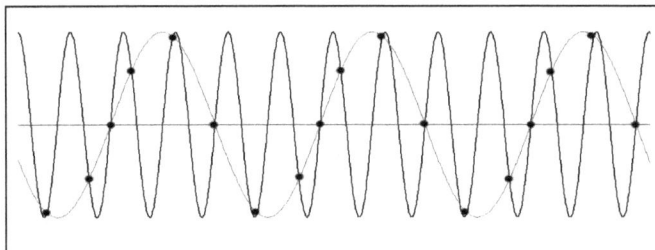

Figure 5.2: Under-sampling giving rise to aliasing

In Figure 5.2 you can see the effect of under-sampling, the low frequency component which shouldn't be there will appear in the sampled data. You very often see this effect in old movies when wheels on horse drawn carriages appear to be rotating backwards. In digital sampling it gives rise to spurious low frequency components as you can see in Figure 5.2. To prevent aliasing from occurring the highest frequency present in the input analogue signal must not exceed twice the sampling rate of the ADC - this is known as the *sampling theorem*. Which basically means that at least two samples are required for the highest frequency present in a sampled waveform as shown in the figure on the left. Often a low pass filter is used to remove frequencies beyond $f_s/2$. Now you have discrete samples to process, the mathematics becomes quite different; move away from continuous functions to discrete values.

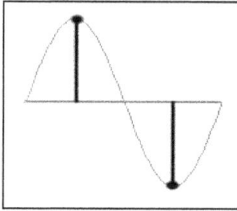

5.2 Difference Equations

When dealing with discrete samples - those which emerge from an ADC the mathematics you need is referred to as *difference equations*. Instead of a continuous variable $x(t)$ you replace it with discrete samples $x(n)$ where n is time index. If the sample period between each sample is Δ then,

$$x(n) = x(t)_{t=n\Delta} \qquad \qquad \dots 5.3$$

This can be visualised as shown in Figure 5.3.

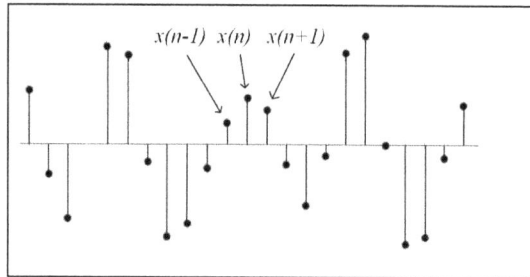

Figure 5.3: For digital samples, n is used as a time index

You will observe that n can start anywhere which gives rise to

$$\dots n+1, n+2, n+3 \dots$$

also

$$\dots n-1, n-2, n-3 \dots$$

In effect difference equations replace differential equations which is required whenever differential equations are implemented on computers in a programming language such as C++ or FORTRAN.

Memory

Memory in a DSP is represented by the symbol shown in Figure 5.4. Difference equations require a means of retaining older samples. As a new sample *x(n)* comes into the memory element D (which is similar to a shift register) and the old sample *x(n-1)* is ejected. The letter D stands for *delay*.

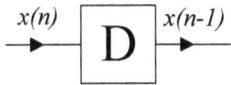

Figure 5.4: A memory element

Derivatives

Given a continuous variable *x(t)* its derivative is $\frac{dx(t)}{dt}$. In a difference equation the first derivative is expressed as,

$$\frac{dx(t)}{dt} \rightarrow \frac{x(n)-x(n-1)}{\Delta} \qquad \ldots 5.4$$

The second derivative is,

$$\frac{d^2x(t)}{dt^2} = \frac{d}{dt}\left(\frac{dx(t)}{dt}\right) \rightarrow \frac{d}{dt}\left(\frac{x(n)-x(n-1)}{\Delta}\right) \rightarrow \frac{x(n)-2x(n-1)+x(n-2)}{\Delta^2} \qquad \ldots 5.5$$

To illustrate how to convert a differential equation into a difference equation we shall consider the following expression for exponential decay,

$$\frac{dx(t)}{dt} = -\lambda x(t) \qquad \ldots 5.6$$

Express this as a difference equation according to Eq:5.4,

$$\frac{x(n)-x(n-1)}{\Delta} = -\lambda x(n) \qquad \ldots 5.7$$

Rearranging this expression,

$$x(n) = \frac{x(n-1)}{1+\lambda\Delta} \qquad \ldots 5.8$$

As an example, substitute 0.5 for Δ, 2 for λ and *x(0)* = 1, then

$$x(n) = \tfrac{1}{2}x(n-1) \qquad \ldots 5.9$$

Plotting this difference equation, starting with an initial value of $x(0) = 1$, you obtain a graph as shown in Figure 5.5.

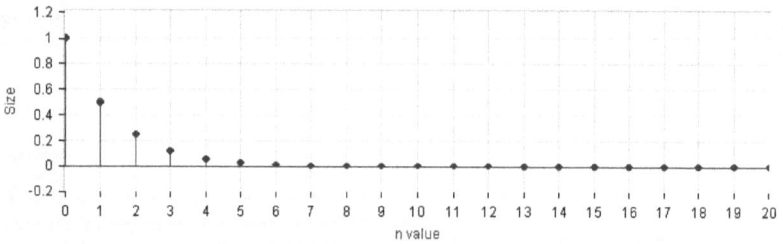

Figure 5.5: Plotting the difference equation for exponential decay

Two other processes can be considered at this stage, differentiation and integration. A differentiation process can be expressed as,

$$y(t) = \frac{dx(t)}{dt} \qquad \text{... 5.10}$$

Using the results in Eq:5.4,

$$y(n) = \frac{x(n) - x(n-1)}{\Delta} \qquad \text{... 5.11}$$

If we let $\alpha = 1/\Delta$, then the digital derivative is x is,

$$y(n) = \alpha[x(n) - x(n-1)] \qquad \text{... 5.12}$$

In a similar manner, for integration if,

$$y = \int x(t)dt \text{ and } \frac{dy}{dt} = x(t) \qquad \text{... 5.13}$$

Expressing this as a difference equation,

$$\frac{y(n) - y(n-1)}{\Delta} = x(n) \qquad \text{... 5.14}$$

Leaving

$$y(n) = \Delta x(n) + y(n-1) \qquad \text{... 5.15}$$

Eq: 5.12 is a *differentiator* and Eq:5.15 is an *integrator*. Digital signal processes are often expressed as difference equations and it is customary to code these processes in a DSPs' assembly language. Every DSP has its own assembly language and an experienced engineer working in the field will have gained a working knowledge of a processor's language. We can now progress to see how difference equations can be used to express processes which we have already covered.

5.3 Digitised the Convolution Function

The convolution integral for a finite duration signal is expressed as,

$$y(t) = \frac{1}{T} \int_0^T h(t') \, x(t' - t) dt' \qquad \ldots 5.16$$

To construct a digital equivalent of Eq:5.16, make the following substitutions,

$$t' \to n, \ t \to m, \ T = N\Delta, \ dt \to \Delta$$

and the integration sign is replaced by a summation sign. The digital convolution becomes,

$$y(m) = \frac{1}{N\Delta} \sum_{n=0}^{n=N-1} h(n) \, x(m - n)\Delta$$

Δ will cancel out leaving,

$$y(m) = \frac{1}{N} \sum_{n=0}^{n=N-1} h(n) \, x(m - n) = h \star x \qquad \ldots 5.17$$

which is the digitised version of the convolution process. N is the number of data samples in the sampled batch over which the convolution will be performed. The summation goes from $n = 0$ to $n = N$-1, since there are N samples, the first one is allocated the index $n = 0$ and the last N-1. Eq:5.17 can be coded directly provided there is a data set representing the impulse response $\{h(n)\}$ for the required process. In practise, when calculating $\{y(m)\}$, $\{h(n)\}$ and $\{x(n)\}$ are transformed into the frequency domain since,

$$h \star x \rightleftharpoons H(k) \cdot X(k) \qquad \ldots 5.18$$

Convolution in the time domain becomes multiplication in the frequency domain which is much easier and quicker to perform.

Moving Average
When considering Eq:5.17, if each stored sample has equal weighting, that is $h(n) = 1$, the expression gives rise to a special case known as a *moving average*,

$$y(m) = \frac{1}{N} \sum_{n=0}^{n=N-1} x(m - n) \qquad \ldots 5.19$$

To appreciate how this performs consider the case when N = 4, then,

$$y(m) = \tfrac{1}{4}[x(m) + x(m-1) + x(m-2) + x(m-3)] \qquad \dots 5.20$$

Now consider the digitised waveform in Figure 5.6.

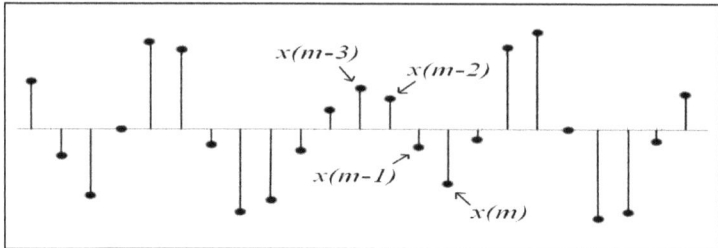

Figure 5.6: Samples contributing to a moving average

The value of *y(m)* is the average of the current value input *x(m)* and the three previous values of *x*. In real-time this moving average would take four samples, calculate the average, then move alone one and calculate the next average. You can think of the moving average process moving along the waveform producing an average of a section of the waveform. In principle you can have as many samples in the process as is required. The moving average can be demonstrated in DADiSP, first construct a waveform with added noise as shown in Figure 5.7.

W1: gsin(300,.01,2)+gnormal(300,.01,0,.3)

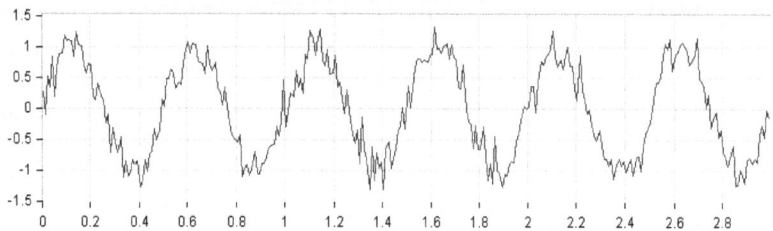

Figure 5.7: A waveform of a noisy signal

Now perform a moving average on the waveform shown in Figure 5.7; the average of 6 samples to produce Figure 5.8.

W2: movavg(w1,6)

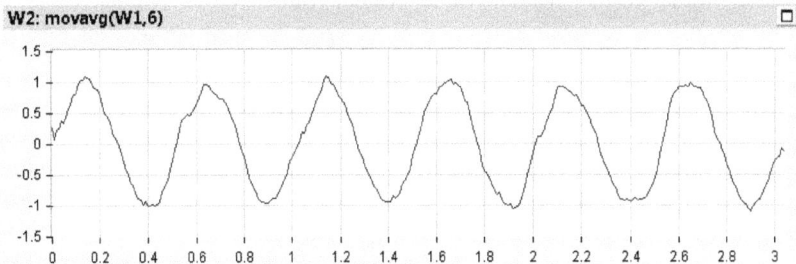

Figure 5.8: The effect of performing a moving average on the
waveform shown in Figure 5.7

You will observe the waveform has been smoothed out considerably. The moving average has the effect of reducing noise in a waveform. Let us now view the spectra before and after the application of the moving average.

| W3: spectrum(w1) |

Now compare the spectra of the waveforms in Figure 5.7 and Figure 5.8 - this is shown in Figure 5.10.

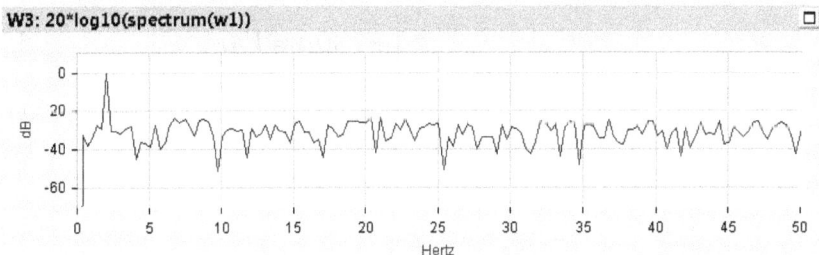

Figure 5.9: The spectrum of the waveform shown in Figure 5.6

| W4: spectrum(w2) |

You will observe from the waveform shown in Figure 5.10 that most of the spectral noise is now below -40dB. Whereas in Figure 5.9 most of the spectral noise is well above -40dB. This clearly demonstrates a reduction in the signal noise. The moving average is often used to provide a real-time view of a signals' trend especially for low frequency waveforms < 1 Hz.

W4: 20*log10(spectrum(w2))

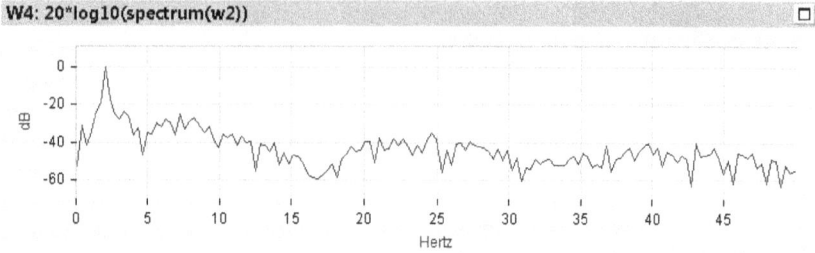

Figure 5.10: Spectrum of waveform shown in Figure 5.7

5.4 A Digital Low Pass Filter

A simple low pass filter is shown in Figure 5.11. It comprises a capacitor C and a resistor R. The output voltage V_o in terms of the input voltage V_i is given by the expression,

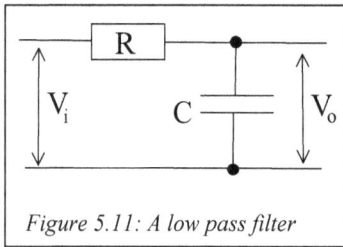

Figure 5.11: A low pass filter

$$RC\frac{dV_o}{dt} + V_o = V_i \qquad ... 5.20$$

This is a first order differential equation and can be solved without too much difficulty. To convert Eq:4.20 into a difference equation, use Eq:5.4 in which case,

$$RC\left[\frac{V_0(n)-V_0(n-1)}{\Delta}\right] + V_0(n) = V_i(n) \qquad ... 5.21$$

This expression can be rearranged to give,

$$V_0(n) = \alpha V_i(n) + \beta V_0(n-1) \qquad ... 5.22$$

where,

$$\alpha = \frac{\Delta}{RC+\Delta} \text{ and } \beta = \frac{RC}{RC+\Delta} \qquad ... 5.23$$

The difference equation in E:5.22 can be coded and implemented directly on a digital signal processor or in a high level language such as C++. We can also observe the effect of this filter by using DADiSP. We can use some example values, α for 0.166 and β for of 0.833. First, create an impulse input.

W1: gimpulse(300,.005,0)

Now implement Eq:5.22 using the filteq command,

W2: Filteq({0.166}, {1, -0.833}, w1)

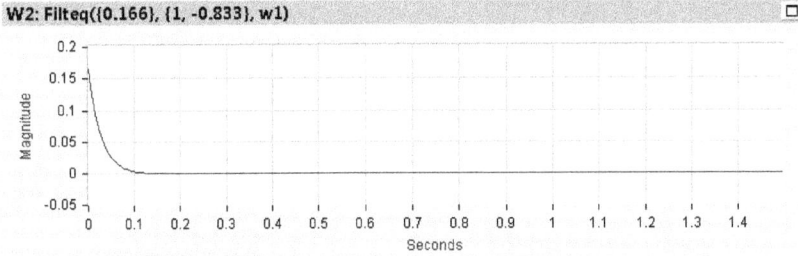

Figure 5.12: The impulse response of Eq:5.22

As observed from Figure 5.12 the output is a broadened impulse and its spectrum is shown in Figure 5.13.

W3: spectrum(w2)

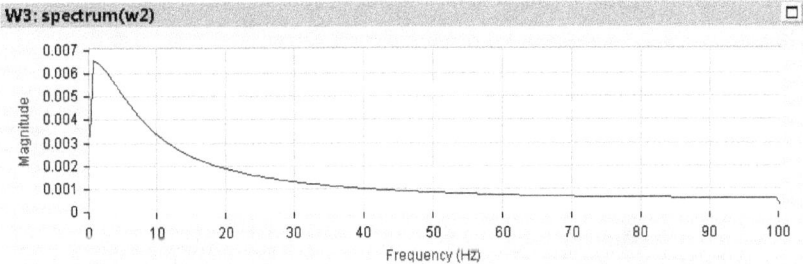

Figure 5.13: Spectrum of the signal shown in Figure 5.12

You will observe the profile in Figure 5.13 displays the characteristics of a low pass filter. A filter comprising a single resistor and capacitor is not well known for having an impressive performance and Figure 5.13 demonstrates this fact. However the exercise shows how a difference equation can be modelled in DADiSP.

5.5 Schematic of a Digital Filter

Eq:5.22 represents a digital filter and you will notice the current output is not only dependent on the current input but also the previous output. This is an example of a *recursive filter* - the delayed output feeds back into the filter. This can be represented as a schematic and is shown in Figure 5.14.

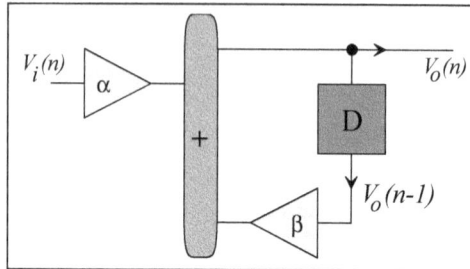

Figure 5.14: A schematic of Eq: 5.22

You will observe from Figure 5.14 two multipliers (for α and β), an adder and a delay D. You will also recollect earlier in this chapter a discussion on the fundamental operations of a Digital Signal Processor (multiplication, addition and memory delay) which you can see in the schematic. In understanding the schematics, note the following steps.

1. The input signal $V_i(n)$ is weighted (multiplied) by α.
2. The input on the left side of the adder is $\alpha V_i(n)$.
3. The output $V_o(n)$ passes through a delay element D.
4. The output from the delay is $V_o(n-1)$ and is then weighted by β to give $\beta V_o(n-1)$.
5. $\beta V_o(n-1)$ passes into the adder and is added to $\alpha Vi(n)$ to produce the output $V_o(n)$.

From following the above steps you will appreciate how the schematic is representative of the digital process. This is especially true for finite impulse response (FIR) filters and infinite impulse response (IIR) filters. Schematics are a very important method for understanding the operation of digital filters and you will encounter them in later chapters.

5.6 Digital to Analogue Conversion

Having looked in some detail at the process of analogue to digital conversion, it's time review the reverse process. Digital to analogue converters (DACs) are intrinsically simpler devices to fabricate. However to ensure their effectiveness an understanding is required of their operation. A certain amount of post processing is required on the signal which emerges from a DAC before an acceptable analogue waveform is produced. It is customary to have an *anti-aliasing filter* on the input to a ADC, in a like manner it's also customary to have a matching anti-aliasing filter on the output of a DAC. A typical output from a DAC is shown in Figure 4.15 which appears as a sample and hold output. As you observe from this figure the output from this DAC is a series of steps. However what is required is a smooth waveform also shown in Figure 4.15. The task in hand is therefore to convert the stepped output from the DAC into the required smoothed waveform. It is in this context that you will very often come across the phrase of *reconstruction filter* and *interpolation filters.* In effect reconstructing the analogue waveform from a sequence of binary numbers. Interpolation is the process of filling in gaps in a waveform from knowledge of the waveform before and after the gaps. If the binary numbers are fed into the DAC at a rate of *fs*, the reconstructed waveform should be band limited to *fs/2*. It is constructive to think of a DAC as an impulse converter.

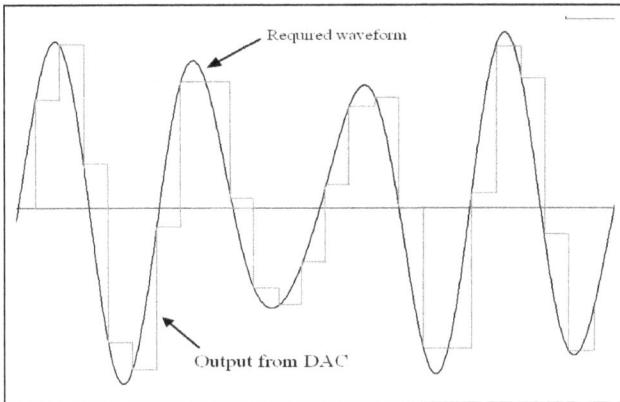

Figure 5.15: Typical output from a DAC

Each binary number fed into the DAC is an impulse, the sequence of impulses of differing sizes forms the profile of the required analogue signal. We can therefore say that,

$$x_s(t) = \sum_{n=-\infty}^{n=+\infty} x(n\Delta)\, \delta(t - n\Delta) \qquad \dots 5.24$$

This expression tells us the analogue $x_s(t)$ is made up of the sequence of digital samples $x(n\Delta)$, one sample produced every Δ seconds (the sampling period). $x_s(t)$ now passes into a low pass filter whose cut-off frequency is half the sampling frequency. The impulse response of an ideal low pass filter for this application is,

$$h(t) = \Delta \frac{\sin(\pi t/\Delta)}{\pi t} \qquad \ldots 5.25$$

which you will recognises as a sinc function. The actual output $y(t)$ from the filter is given by,

$$y(t) = \int_{-\infty}^{\infty} h(t - \tau)\, x_s(\tau) d\tau \qquad \ldots 5.26$$

Substituting Eq:5.25 and Eq:5.26 it can be shown that,

$$y(t) = \sum_{n=-\infty}^{n=+\infty} x(n\Delta) \frac{\sin[\pi(t-n\Delta)/\Delta]}{\pi(t-n\Delta)/\Delta} \qquad \ldots 5.27$$

This expression indicates that every digital sample emerging from the DAC comes out as a sinc function as shown in Figure 14.16. The actual analogue waveform from the DAC follows the peaks of the sinc functions and the rest of the sinc profile *fills* in the gaps between the peaks - *interpolating between the peaks.*

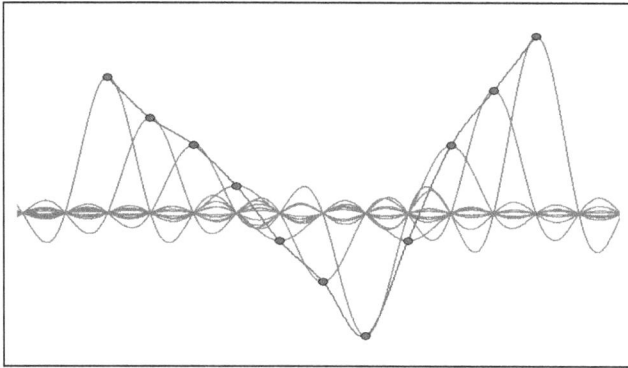

Figure 5.16: Each output from the DAC creates a sinc function

Although it may not appear very convincing from Figure 5.16, it is nonetheless a very effective process for producing a band limited analogue signal.

What you have gained from this Chapter
1. An understanding of an algorithm.
2. The concept of digitising analogue signals.
3. The effect of aliasing on the digitised waveform.
4. How difference equations are used to represent differential equations and discrete samples.
5. How the convolution function is digitised.
6. An understanding of the moving average.
7. How a low pass digital filter is derived from an analogue filter.
8. Representing a digital filter with a schematic.
9. An insight into the concept of digital to analogue conversion.

DADiSP skills you acquired from this Chapter
1. *gsin* - generating a sine wave
2. *gnormal* - generating a random waveform which has a known variance and mean.
3. *movavg* - implementing a moving average process on a waveform.
4. *gimpulse* - creating an impulse
5. *filteq* - creating a digital filter using coefficients
6. *spectrum* - obtaining a spectrum of a waveform

DADiSP Extra

To create a square wave with added noise, enter the following commands, the resulting waveform should be similar to Figure 5.17.

W2: gsqrwave(300, .01, 2.0, 0)+gnormal(300,.01,0,.1)

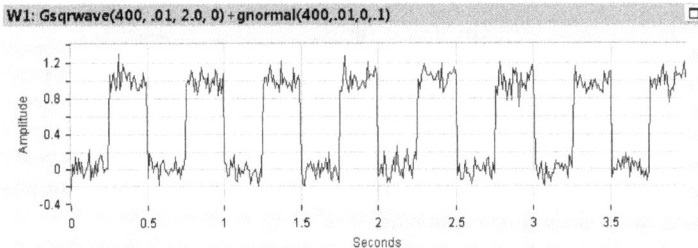

Figure 5.17: A noisy square wave

71

6. The Sinc Function

A function which is extremely important in the field of digital signal processing is the sinc function and this is another example of a *generalised function*. It is defined as,

$$\mathrm{sinc}(x) = \frac{\sin(x)}{x} \qquad \ldots 6.1$$

You will observe from this function that when $x = 0$ the result is $\frac{0}{0}$, which is an *indeterminate number* - it can be anything! By convention sinc(0) =1. There is a command in DADiSP for generating a sinc function which can be seen in Figure 6.1.

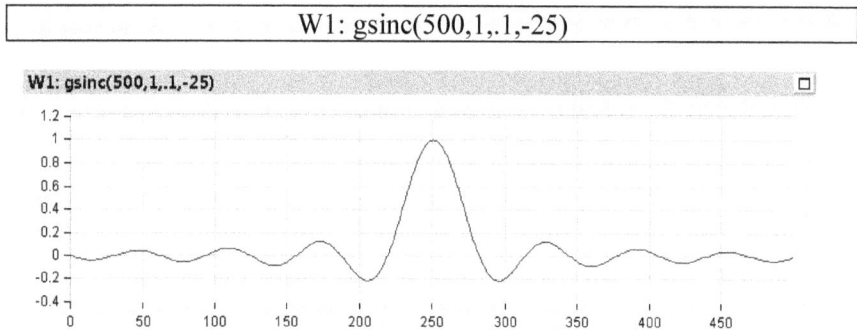

W1: gsinc(500,1,.1,-25)

Figure 6.1: The sinc function

The central peak is referred to as the *main lobe* and on either side of it are the *side lobes*. You will observe the function ripples along the zero x-axis which indicates there are several values of x for which sinc(x) = 0. In fact,

$$\mathrm{sinc}(x) = 0\big|_{x=n\pi} \qquad \ldots 6.2$$

The zero values correspond to the same values for when sin(x) = 0 where x is expressed in radians. When discussing *Finite Impulse Response* (FIR) filters an understanding of the sinc function is crucial and this is covered in greater depth in Chapter 8. It is interesting to note the sinc function can be compressed as shown in Figure 6.2.

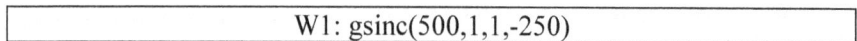

W1: gsinc(500,1,1,-250)

W1: gsinc(500,1,1,-250)

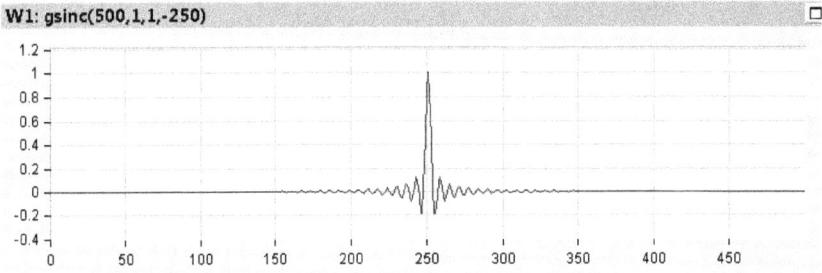

Figure 6.2: A compressed sinc function

Alternatively the sinc function can be broadened as shown in Figure 6.3

W1: gsinc(500,1,.06,-15)

W1: gsinc(500,1,.06,-15)

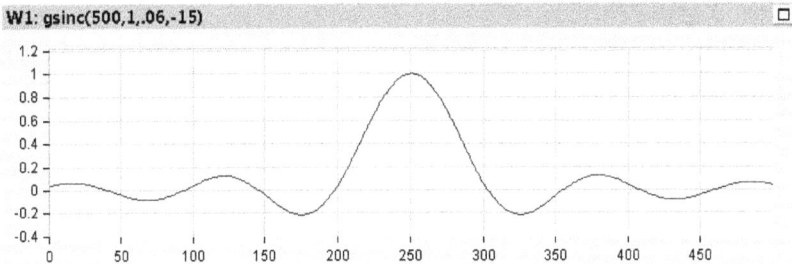

Figure 6.3: A broadened sinc function

The shape of the sinc profile is very significant which will become apparent later. It's worth mentioning that Figure 6.2 shows the sinc function becoming very narrow. In fact a property of the sinc function is,

$$\underset{a \to 0}{Lim} \left[\tfrac{1}{a}\text{sinc}(\tfrac{x}{a}) \right] = \delta(x) \qquad \text{... 6.3}$$

This expression states that as the sinc function becomes narrower and narrower, in the *limit* it becomes a *Dirac Impulse Function* which was discussed in Chapter 4.

6.1 Euler's Formula
Before we can proceed it is necessary to discuss what is referred to as *Euler's Formula* which is a method of representing sine and cosine in

complex notation. Euler was a Swiss mathematician who made several important contributions to mathematics especially in the field of *analysis* which include the *theory of functions*. Euler's formula is expressed as,

$$e^{ja\theta} = \cos(a\theta) + j\sin(a\theta) \quad \text{... 6.4}$$

There are two special values of α of interest, when $\alpha = 1$, giving,

$$e^{j\theta} = \cos(\theta) + j\sin(\theta) \quad \text{... 6.5}$$

Leonhard Euler
1707 - 1783

and when $\alpha = -1$, giving

$$e^{-j\theta} = \cos(\theta) - j\sin(\theta) \quad \text{... 6.6}$$

This occurs because $sin(-\theta) = -sin(\theta)$ and $cos(-\theta) = cos(\theta)$. When we add Eq:6.5 and Eq:6.6 we get,

$$e^{j\theta} + e^{-j\theta} = 2\cos(a\theta) \quad \text{... 6.7}$$

Leaving,

$$\cos(\theta) = \frac{e^{j\theta} + e^{-j\theta}}{2} \quad \text{... 6.8}$$

And when we subtract Eq:6.6 from Eq:6.5 we get,

$$e^{j\theta} - e^{-j\theta} = 2j\sin(\theta) \quad \text{... 6.9}$$

Leaving

$$\sin(\theta) = \frac{e^{j\theta} - e^{-j\theta}}{2j} \quad \text{... 6.10}$$

Eq:6.8 and 6.10 are encountered on many occasions in digital signal processing and it's a good idea to commit these to memory. Should you feel inclined, you can expand these two equations as series and confirm their identity. Also from Eq:6.4,

$$e^{ja\theta} = (e^{j\theta})^a = [\cos(\theta) + j\sin(\theta)]^a \quad \text{... 6.11}$$

$$= \cos(a\theta) + j\sin(a\theta) \quad \text{... 6.12}$$

where α usually takes on a real value. This is known as *de Moivre's Theorem* named after *Abraham de Moivre (1667 - 1754)* the French mathematician.

6.2 Fourier Transform of a Sinc Function

Having defined a sinc function, the next task is to examine its profile in the frequency domain - to find its spectrum. To achieve this, we first perform a Fourier Transform on a rectangular pulse as shown in Figure 6.4.

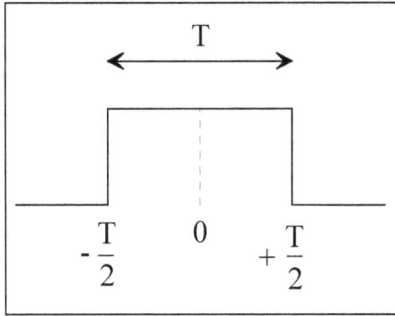

Figure 6.4: A rectangular pulse of duration T seconds

The amplitude of the pulse is 1 unit, the Fourier Transform is therefore expressed as,

$$F(\omega) = \int_{-T/2}^{T/2} 1e^{-j\omega t}dt \qquad \ldots 6.13$$

This can be integrated directly,

$$F(\omega) = \left[\frac{e^{-j\omega t}}{-j\omega} \right]_{-T/2}^{T/2} \qquad \ldots 6.14$$

This becomes,

$$F(\omega) = \frac{1}{j\omega}\left[e^{j\frac{\omega T}{2}} - e^{-j\frac{\omega T}{2}} \right]$$

$$= \frac{T}{2}\left[\frac{e^{j\frac{\omega T}{2}} - e^{-j\frac{\omega T}{2}}}{j\frac{\omega T}{2}} \right] \qquad \ldots 6.15$$

Using Eq:6.11 we obtain,

$$F(\omega) = T\frac{\sin(\frac{\omega T}{2})}{\frac{\omega T}{2}} = T\operatorname{sinc}(\frac{\omega T}{2}) \qquad \ldots 6.16$$

The Fourier Transform of a rectangular pulse is therefore a sinc function, this is a very important result and its significance will become evident in Chapter 8.

6.3   Features of the Sinc Function

Figure 6.5 shows a cosine wave which is limited in time - a duration of T seconds - which corresponds to the length of the box. In the previous section we have seen what happens when a Fourier Transform is performed on a square (or rectangular) pulse. It is now constructive to think of Figure 6.5 as rectangular pulse with a cosine wave of frequency ω_0 contained within it.

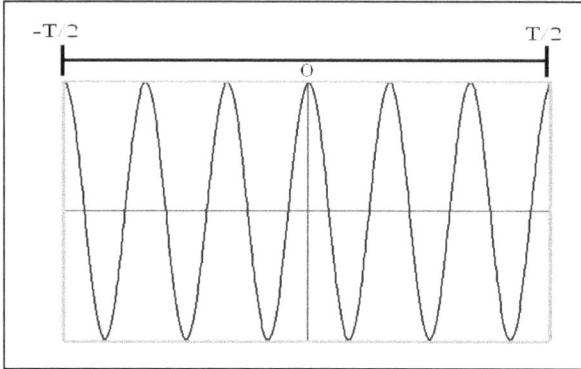

Figure 6.5: A cosine wave of limited duration

The Fourier Transform of this waveform is given by,

$$F(\omega) = \int_{-T/2}^{T/2} \cos(\omega_0 t)\, e^{-j\omega t} dt \qquad \text{... 6.17}$$

Use Eq:6.8,

$$F(\omega) = \frac{1}{2} \int_{-T/2}^{T/2} [e^{j\omega_0 t} + e^{-j\omega_0 t}]\, e^{-j\omega t} dt \qquad \text{... 6.18}$$

which becomes,

$$F(\omega) = \frac{1}{2} \int_{-T/2}^{T/2} [e^{j(\omega_0-\omega)t} + e^{-j(\omega_0+\omega)t}] dt \qquad \text{... 6.19}$$

Perform the integration,

$$F(\omega) = \frac{1}{2}\left[\frac{e^{j(\omega_0-\omega)t}}{j(\omega_0-\omega)} \right]_{-T/2}^{T/2} + \frac{1}{2}\left[\frac{e^{-j(\omega_0+\omega)t}}{j(\omega_0+\omega)} \right]_{-T/2}^{T/2} \qquad \text{... 6.20}$$

Expanding the brackets gives,

76

$$F(\omega) = \frac{1}{2j(\omega_0 - \omega)}\left[e^{j(\omega_0 - \omega)\frac{T}{2}} - e^{-j(\omega_0 - \omega)\frac{T}{2}}\right]$$

$$+ \frac{1}{2j(\omega_0 + \omega)}\left[e^{-j(\omega_0 + \omega)\frac{T}{2}} - e^{j(\omega_0 + \omega)\frac{T}{2}}\right] \qquad \dots 6.21$$

which becomes,

$$F(\omega) = T\left[\frac{e^{j(\omega - \omega_0)\frac{T}{2}} - e^{-j(\omega - \omega_0)\frac{T}{2}}}{j(\omega - \omega_0)\frac{T}{2}}\right] + T\left[\frac{e^{j(\omega + \omega_0)\frac{T}{2}} - e^{-j(\omega + \omega_0)\frac{T}{2}}}{j(\omega - \omega_0)\frac{T}{2}}\right] \qquad \dots 6.22$$

Using Eq:6.10,

$$F(\omega) = T\operatorname{sinc}(\omega - \omega_0)\frac{T}{2} + T\operatorname{sinc}(\omega + \omega_0)\frac{T}{2} \qquad \dots 6.23$$

Eq:6.23 shows two sinc functions, one with frequency $(-\omega_0)$ and the other with frequency $(+\omega_0)$ and Figure 6.6 shows the frequency profiles of the sinc functions.

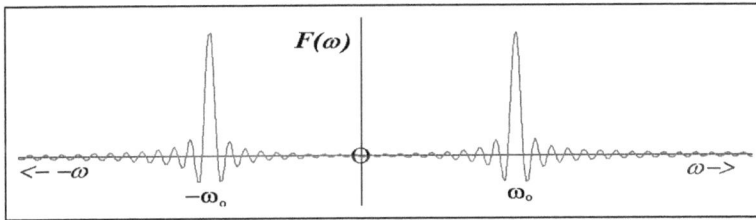

Figure 6.6: The spectral profiles from truncated sine wave

Figure 6.6 illustrates that when a cosine wave is truncated in time it produces two sinc functions in the frequency domain. One in the positive half of the spectrum and one in the negative half. As the duration of the cosine wave gets larger Eq:6.3 comes into play which produces two spectral lines. With a truncated sine wave the energy leaks into the *sidelobes* of the sinc function and this is referred to as *spectral leakage* and prevents an accurate measurement of magnitude being performed on the spectral lines. Referring to Figure 6.5, the sinc function itself comes from the rectangular profile and the position of the sinc profile from the frequency of the cosine wave.

You may want to consider Eq:6.16 as the spectral profile of DC (0 Hz). As the duration of the pulse gets longer, the sinc profile would narrow down to a single line at 0 Hz. Similar to measuring the voltage on a battery, taking time T to perform the measurement and then disconnecting from it. We can now use DADiSP to simulate the behaviour of the sinc function, first create a cosine wave.

W1: gcos(200,.01,9)

Now create a line

W2: gline(20,.01,0,0)

Concatenate them as shown in Figure 6.7,

W3: concat(w2,w1,w2)

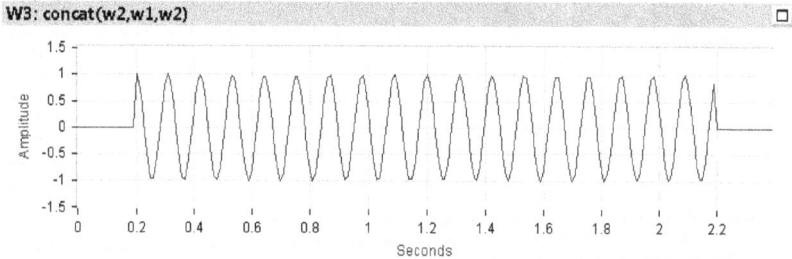

Figure 6.7: A truncated cosine wave

You will observe from Figure 6.7, the lines (zero values) at the start and end of the cosine wave, in effect limiting the duration of the cosine wave - referred to as a *truncated* waveform. The spectrum of a truncated cosine wave has some special features, to obtain the spectrum of the truncated wave use the following command - the results are shown in Figure 6.8.

W4: spectrum(w3)

You will notice in Figure 6.8 the presence of side lobes around the spectral line although not as prominent as shown Figure 6.6. It should be noted the *y*-axis of Figure 6.8 is magnitude which is all positive. In reality when performing spectral analysis efforts are made to minimise the magnitude of the side-lobes around a spectral line and this theme will be discussed in greater detail in Chapter 11.

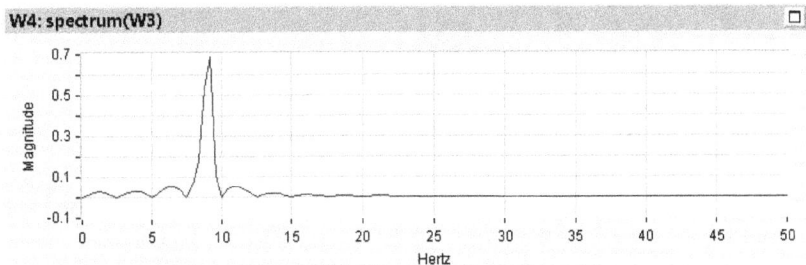

Figure 6.8: Spectrum of waveform shown in Figure 6.7

What you have gained from this chapter
1. How a sinc function is defined.
2. How a sinc function narrows in the limit to a Dirac Impulse function.
3. The significance of Euler's equations.
4. Making sense of spectral profiles in the context of sinc functions.
5. Appreciating the reason for spectral leakage.

DADiSP skills you have acquired from this Chapter
1. *gsin* - generate a sine wave.
2. *gline* - to generate a line
3. *concat* - joining together different waveforms
4. *spectrum* - obtaining a spectrum of a waveform

7. The Z-transform

To analyse the performance of digital filters it's necessary to have a reasonable knowledge of Z-transforms. The Z-transform transforms data from the discrete time domain into the Z domain which provides insight into the behaviour of digital filters. Given a batch of samples which have emerged from an ADC, the Z-transform is given by,

$$Z\{x(n)\} = \sum_{k=0}^{k=+\infty} x(k)z^{-k} = X(z) \qquad \ldots 7.1$$

As it stands this is not very useful as an engineer never deals with an infinite number of samples. There are a few conditions relating to Z-transforms of which you should be aware. The first is the *linearity condition*,

$$\begin{aligned} Z\{ax(n) + by(n)\} &= aZ\{x(n)\} + bZ\{y(n)\} \\ &= a\,X(z) + b\,Y(z) \end{aligned} \qquad \ldots 7.2$$

The second is the *shift property,*

$$Z\{x(n-k)\} = X(z)z^{-k} \qquad \ldots 7.3$$

As you can see in Eq:7.2 and Eq:7.3 the Z-transform performs on discrete data only. We can apply the Z-transform directly on a digital filter. Consider the following second order process,

$$y(n) = b_0 x(n) + b_1 x(n-1) + b_2 x(n-2) \qquad \ldots 7.4$$

The Z-transform of this expression becomes,

$$Z\{y(n)\} = b_0 Z\{x(n)\} + b_1 Z\{x(n-1)\} + b_2 Z\{x(n-2)\} \qquad \ldots 7.5$$

Using the shift theorem,

$$\begin{aligned} Y(z) &= b_0 X(z) + b_1 X(z)z^{-1} + b_2 X(z)z^{-2} \\ &= X(z)[b_0 + b_1 z^{-1} + b_2 z^{-2}] \end{aligned} \qquad \ldots 7.6$$

The *transfer function* for the filter is defined as,

$$H(z) = \tfrac{Y(z)}{X(z)} = [b_0 + b_1 z^{-1} + b_2 z^{-2}] \qquad \ldots 7.7$$

Eq:7.7 can be expressed as,

$$H(z) = \frac{b_0 z^2 + b_1 z + b_2}{z^2} \qquad \qquad ... 7.8$$

You will observe the numerator in Eq:7.8 is a quadratic which can be factorised. If z_1 and z_2 are the roots of the quadratic, then

$$H(z) = \frac{(z - z_1)(z - z_2)}{z^2} \qquad \qquad ... 7.9$$

where the roots are,

$$z_1 = \frac{-b_1 + \sqrt{b_1^2 - 4b_0 b_2}\,)}{2b_0}$$
$$z_2 = \frac{-b_1 - \sqrt{b_1^2 - 4b_0 b_2}\,)}{2b_0} \qquad \qquad ... 7.10$$

7.1 The Concept of the Zero

You will note from Eq:7.9 that $H(z) = 0$ when $z = z_1$ or when $z = z_2$. These values are called *zeros* - since the transfer function $H(z)$ is zero at these points. Sometimes the roots are complex - having real and imaginary components. Eq:7.10 can be written in a complex format,

$$z_1 = -\frac{b_1}{2b_0} + j\frac{\sqrt{4b_0 b_2 - b_1^2}\,)}{2b_0}$$
$$z_2 = -\frac{b_1}{2b_0} - j\frac{\sqrt{4b_0 b_2 - b_1^2}\,)}{2b_0} \qquad \qquad ... 7.11$$

You will notice from Eq:7.11 the zero values occur in conjugate pairs in which case,

$$z_1 = z_2^* \qquad \qquad ... 7.12$$

In another illustrative example we can consider the process,

$$y(n) = x(n) - x(n - 4) \qquad \qquad ... 7.13$$

When a Z-transform is performed on this expression we get,

$$Z\{y(n)\} = Z\{x(n)\} - Z\{x(n - 4)\} \qquad \qquad ... 7.14$$

which becomes

$$Y(z) = X(z) - X(z) z^{-4} \qquad ...\,7.15$$

The transfer function is,

$$H(z) = 1 - z^{-4} = \frac{z^4-1}{z^4} \qquad ...\,7.16$$

The nominator can be factorised to give,

$$H(z) = \frac{(z+1)(z-1)(z+j)(z-j)}{z^4} \qquad ...\,7.17$$

You will observe from Eq:7.17 there are four zeros in this process, two real and two imaginary. An alternative way of thinking of complex roots is to consider Figure 7.1 which shows plots of

$$y(x) = x^2 - 1 = (x+1)(x-1)$$

and

$$w(x) = x^2 + 1 = (x+j)(x-j).$$

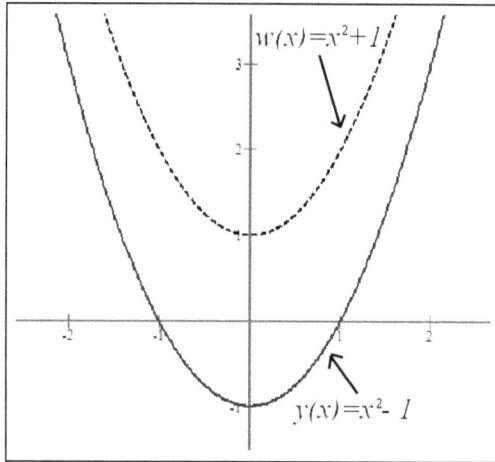

Figure 7.1: Quadratic functions, one with real roots the other with imaginary roots

The roots of *y(x)* are clearly visible as the function passes through the *x*-axis. Whereas *w(x)* is also quadratic, but since the function does not pass through the *x*-axis you can infer its roots are complex. The significance of this fact will become evident in Chapter 8 as you encounter several expressions where the roots are complex.

7.2   The Inverse Z-transform

The formal method of finding the inverse Z-transform (IZT) involves the use of analytical *contour integration*, to be precise,

$$x(n) = \frac{1}{2\pi j} \oint X(z) \, z^{n-1} dz \qquad \text{... 7.18}$$

Fortunately you only need to use this integral on rare occasions. Usually the only process which is normally needed is,

$$x(n - k) = Z^{-1}\{X(z)z^{-k}\} \qquad \text{... 7.19}$$

To appreciate this process, consider an example.

$$Y(z) = X(z) - X(z)z^{-2} \qquad \text{... 7.20}$$

Perform the IZT,

$$Z^{-1}\{Y(z)\} = Z^{-1}\{X(z)\} - Z^{-1}\{X(z)z^{-2}\} \qquad \text{... 7.21}$$

Using Eq:7.19, leaves,

$$y(n) = x(n) - x(n-2) \qquad \text{... 7.22}$$

You will need to use the inverse Z-transform when considering FIR and IIR digital filters.

7.3 Passing from Z to Frequency

Very often having analysed a digital process in the Z domain, the question arises as to how to relate it to its activity in the frequency domain. To convert from the Z domain to the frequency domain, make the following substitution,

$$z = e^{i\omega\Delta} \qquad \text{... 7.23}$$

where Δ is the sampling interval - this remains constant - it is after all a feature of a analogue to digital converter. To demonstrate how this is used, consider the process,

$$y(n) = x(n) + x(n-2) \qquad \text{... 7.24}$$

Perform a Z-transform on this expression,

$$Z\{y(n)\} = Z\{x(n)\} + Z\{x(n-2)\} \qquad \text{... 7.25}$$

which becomes.

$$Y(z) = X(z) + X(z) \, z^{-2} \qquad \text{... 7.26}$$

The transfer function is,

$$H(z) = \frac{z^2+1}{z^2} \qquad \dots 7.27$$

Now make the substitution from Eq:7.23,

$$H(\omega) = \frac{(e^{j2\omega\Delta}+1)}{e^{j2\omega\Delta}} \qquad \dots 7.28$$

The complex conjugate is formed by replacing j with $-j$,

$$H^*(\omega) = \frac{(e^{-j2\omega\Delta}+1)}{e^{-j2\omega\Delta}} \qquad \dots 7.29$$

Therefore,

$$|H(\omega)|^2 = (e^{j2\omega\Delta}+1)(e^{-j2\omega\Delta}-1) = 1 + e^{j2\omega\Delta} + e^{-j\omega\Delta} + 1 \qquad \dots 7.30$$

Leaving,

$$H(\omega) = 2\{1 + \cos(2\omega\Delta)\} \qquad \dots 7.31$$

This is called a *raised cosine* filter and is shown in Figure 7.2,

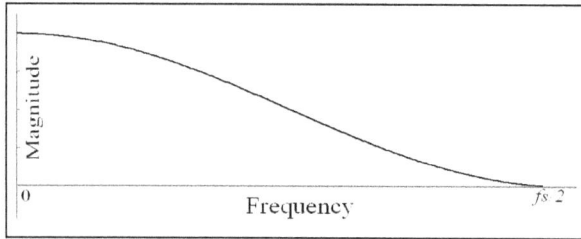

Figure 7.2: A raised cosine profile

From Figure 7.2 you will observe the raised cosine profile exhibits the appearance of a low pass filter. There is a command in DADiSP which converts from the Z-plane into the frequency domain. Consider the expression in the Z-plane

$$H(z) = 1 + 0.9z^{-1} + 0.6z^{-2} - 0.3z^{-3} - 0.5z^{-4} \qquad \dots 7.32$$

The DADiSP command is

W1: zfreq({1, 0.9, 0.6, -0.3, -0.5}, {1}, 1024)

and the result is shown in Figure 7.3,

W2: zfreq({1, 0.9, 0.6, -0.3, -0.5}, {1}, 1024)

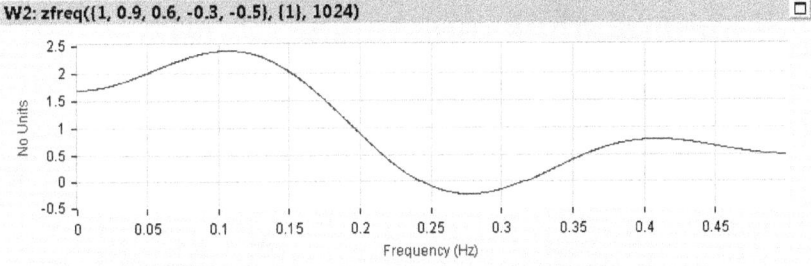

Figure 7.3: The transfer function of Eq:7.32

When you execute the DADiSP command zfreq the result is complex, you are therefore able to determine the magnitude and phase spectra from zfreq.

7.4 Sums and Products

An alternative method of representing higher order processes is often used and we can introduce the idea of a *polynomial*,

$$y(x) = a_0 + a_1x + a_2x^2 + a_3x^3 + ...a_nx^n \qquad ...7.33$$

This is referred to a nth order polynomial as the highest index on x is n. The short hand notation for Eq:7.33 is,

$$y(x) = \sum_{n=0}^{m=n} a_mx^m \qquad ...7.34$$

It was demonstrated in Eq:7.7 that if you have a second order expression, which is equivalent to a second order polynomial, it can be factorised into two roots. Likewise if you have an nth order polynomial it can be factorised into n roots. Eq:7.34 can be therefore be written as,

$$y(x) = (x - x_1)(x - x_2)(x - x_3)...(x - x_n) \qquad ...7.35$$

where $x_1, x_2, ... x_n$ are the roots of $y(x)$. This expression can be written as,

$$y(x) = \prod_{m=1}^{m=n}(x - x_m) \qquad ...7.36$$

The Greek letter Π represents the *product* (capital P) and you can see that the sum in Eq:7.34 is now expressed in terms of the factorised products in Eq:7.36. You will note that if any of the factors in Eq:7.36 are zero then *y(x)* is also zero. Factorising high order expressions (converting Eq:7.34 into Eq:7.36) can only be performed numerically and is no easy task. When discussing digital filters in detail, you will be require an understanding of these sums and products. Having gained an understanding of the concept of zeros and the Z plane, we can now progress to a visual method for representing the behaviour of zeros in what is often referred to as the *Z-plane*. The *unit circle* is an essential tool for understanding the behaviour of digital filters.

7.5 The Unit Circle

Crucial to the understanding of how digital processes function is the concept of the unit circle. This is shown in Figure 7.4. There a number of features of the unit circle which you should appreciate.

- Moving from 0Hz to $f_s/2$ anti-clockwise represents the positive frequency direction.
- Moving from 0Hz to $-f_s/2$ clockwise represents the negative frequency direction.
- The horizontal line drawn from 0 Hz to $f_s/2$ is the *real axis*.
- The vertical line passing through the origin represent the *imaginary axis*.

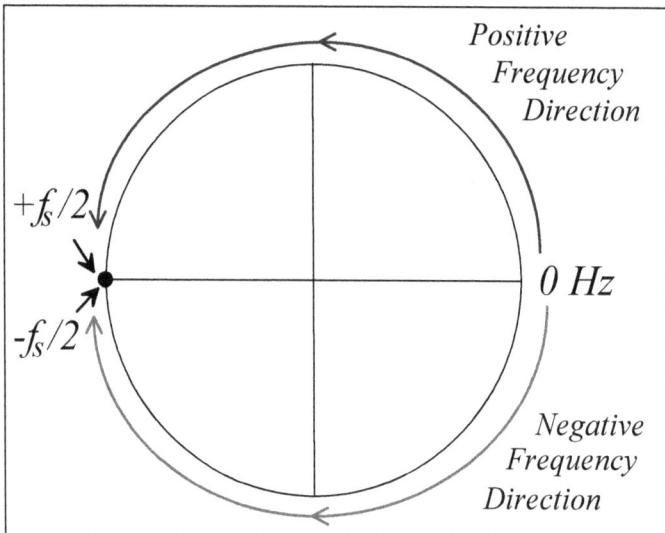

Figure 7.4: the unit circle shows the locations of DC (0 Hz) and $f_s/2$ half the sampling frequency.

The performance of a digital process at any given frequency will depend on the contents of the unit circle and closeness of features to the perimeter of the unit circle. As an example consider Eq:7.16 where there are four zeros. In Figure 7.5 you will observe the actual positions of the zeros in the unit circle. A zero is represented as a little circle - in effect a zero. You will also observe the conjugate position of the zeros in the unit circle - the lower half of the circle is a mirror image of the upper half. An obvious question is how do these zeros affect the performance of the process given in Eq:7.16? From the previous discussion it was pointed out that where the zeros occur, the transfer function $H(z) = 0$. We can use DADiSP to observe the spectral effect of the zeros in Eq:7.17. First generate an impulse.

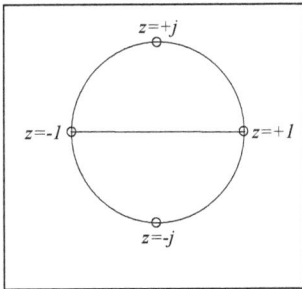

Figure 7.5: Showing the zero values of Eq:7.16

W1: gimpulse(300,.1,0)

This impulse will serve as $x(n)$. Perform a delay of data values to give $x(n-4)$.

W2: delay(w1,4)

Now combine them to give $x(n)-x(n-4)$ as shown in Figure 7.6

W3: w1-w2

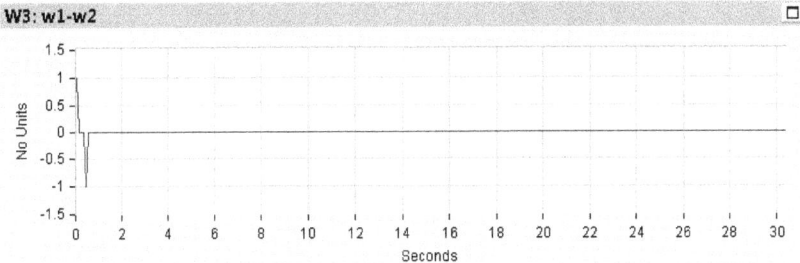

Figure 7.6: Implementation of the expression x(n)-x(n-1)

Now determine the spectrum of this process, as shown in Figure 7.6,

W4: spectrum(w3)

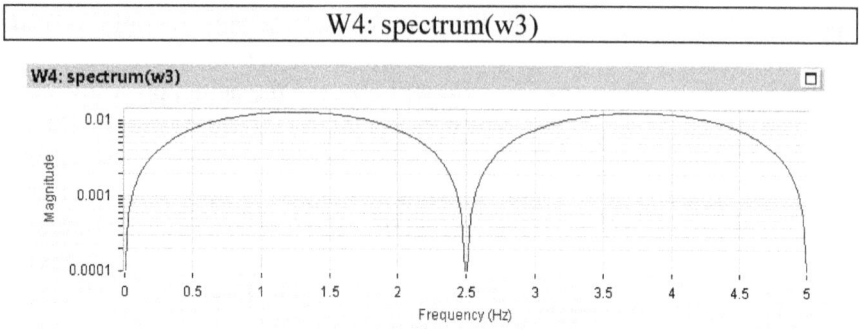

Figure 7.7: This show the effect of the zeros in Eq:7.17.

Now make a comparison of Figure 7.5 with Figure 7.7. You will note that there are three zeros present in Figure 7.7. What you see is the effect of moving anticlockwise starting from DC where there is a zero. As you progress you will encounter another zero midway and when you reach the end you encounter the last zero situated at $f_s/2$. If you progress in a clockwise along the negative frequency you will encounter a similar pattern. Also note that the $f_s/2$ zero coincides with the $-f_s/2$ zero. There is a command in DADiSP for creating the unit circle in the Z-plane. We shall use the following third order difference equation,

$$y(n) = x(n) + 0.7x(n-1) - 0.3x(n-2) - 0.6x(n-3) \qquad \dots 7.37$$

W1: zplane({1, 0.7, -0.3, -0.6}, {1})

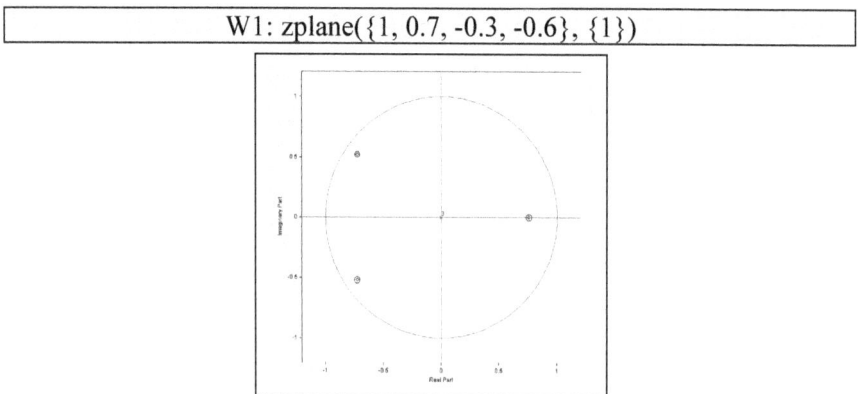

Figure 7.8: The zplane instruction showing the zero locations

You will recognise how the coefficients for the zplane instruction are taken directly from the difference equation Eq:7.37 and three zeros are shown in Figure 7.8. Make a comparison between Figure 7.4 and Figure 7.8. Although the position of 0 Hz and *fs/2* are not marked on Figure 7.8, you should recognise their locations.

7.6 Regions of Convergence

When considering Z-transform it is important to be aware of regions in the Z-plane where there are potential problems. In the definition of the Z-transform, it was assumed that Eq:7.1 converges to a finite value - after all it is an infinite sum. Is it therefore reasonable to make this assumption? Among the tests for convergence is the *Ratio Test* which states that if the ratio of two consecutive values in a series,

$$\left|\frac{x_{n+1}}{x_n}\right| \text{ converges to } L < 1$$

then the series $\sum x_n$ converges absolutely. When applied to sampled data,

$$\underset{n\to\infty}{Lim}\left|\frac{x_{n+1}}{x_n}\right| = \underset{n\to\infty}{Lim}\left|\frac{x(n+1)z^{-n-1}}{x(n)z^{-n}}\right| = \left[\underset{n\to\infty}{Lim}\left|\frac{x(n+1)}{x(n)}\right|\right] \cdot |z|^{-1} \qquad \dots 7.38$$

For convergence of the series then,

$$|z|^{-1} \underset{n\to\infty}{Lim}\left|\frac{x(n+1)}{x(n)}\right| < 1 \qquad \dots 7.39$$

The *Region of Convergence* (ROC) is given by

$$|z| > \underset{n\to\infty}{Lim}\left|\frac{x(n+1)}{x(n)}\right| = R \qquad \dots 7.40$$

This expression tells us the series will converge provided $|z|$ is outside the circle of radius R.

The Unit Step Function

Consider a unit step function (another *generalised function*), as shown in Figure 7.9.

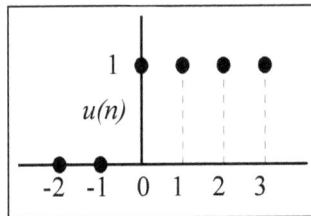

Figure 7.9: The unit step function

The unit step function is defined as,

$$u(n) = \begin{array}{ll} 0 & n < 0 \\ 1 & n \geq 0 \end{array} \qquad \text{... 7.41}$$

The Z-transform of the unit step is,

$$Z\{u(n)\} = \sum_{n=1}^{\infty} u(n)z^{-n} = \sum_{n=1}^{\infty} z^{-n} \qquad \text{... 7.42}$$

When this sum is expanded,

$$Z\{u(n)\} = z^{-1} + z^{-2} + z^{-3} + z^{-4} + ... \qquad \text{... 7.43}$$

This series can be expressed as,

$$Z\{u(n)\} = z^{-1}\left[1 + z^{-1} + z^{-2} + z^{-3} + ...\right] \qquad \text{... 7.44}$$

and the left hand side of this series can be written as,

$$Z\{u(n)\} = \frac{z^{-1}}{1 - z^{-1}} \qquad \text{... 7.45}$$

The ROC therefore occurs when $|z^{-1}| < 1$. As you can appreciate problems occur with convergence of Eq:7.45 if this condition is not met.

7.7 Digitising a Rectangular Pulse

You are already familiar with the sinc function in the frequency domain which has been derived from a pulse in the time domain. When considering a digitised rectangular pulse in the time domain, as shown in Figure 7.10, the result you obtain in the frequency domain is slightly different.

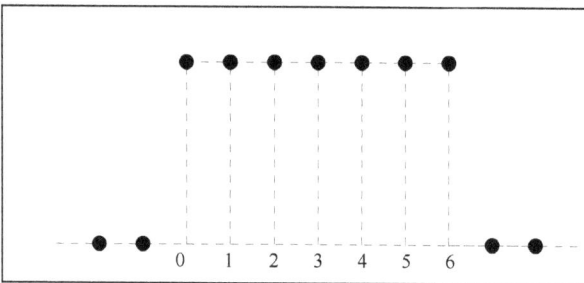

Figure 7.10: A digitised rectangular pulse

The sample values are $\{1,1,1,1,1,1,1\}$ - all other values are zero. When a Z-transform is performed on this data set, we get,

$$Z\{y(t)\} = \sum_{n=0}^{n=7} x(n)z^{-n} = \sum_{n=0}^{n=7} z^{-n} \qquad \text{... 7.46}$$

Expanding the sum,

$$Y(z) = 1 + z^{-1} + z^{-2} + z^{-3} + z^{-4} + z^{-5} + z^{-6} + z^{-7} \qquad \text{... 7.47}$$

This is a geometric series, which means that Eq:7.47 can be expressed as,

$$Y(z) = \frac{1-z^{-8}}{1-z^{-1}} = \frac{z^{-4}}{z^{-\frac{1}{2}}} \left(\frac{z^4 - z^{-4}}{z^{\frac{1}{2}} - z^{-\frac{1}{2}}} \right) \qquad \text{... 7.48}$$

Now let $z = e^{j2\pi\omega T}$, then

$$|Y(\omega)|^2 = \frac{\sin^2(4\omega T)}{\sin^2(\omega T/2)} \qquad \text{... 7.49}$$

This is very similar to what is known as the *Dirichlet Kernel*, which has the form,

$$\frac{\sin((n+0.5)x)}{\sin(x/2)} \qquad \text{... 7.50}$$

Figure 7.11 shows this function with various values x.

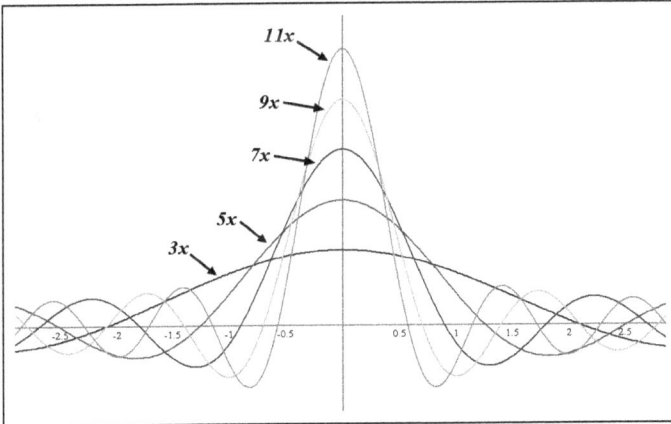

Figure 7.11: The Dirichlet Kernel with various values of x

As the value of x increases the main peak gets sharper and produces more ripples and has a close resemblance to that of the sinc function. The Dirichlet kernel is the transfer function of a *diffraction grating* which is an instrument used for processing optical signals.

Figure 7.12: The effect of optical diffraction from a CD

A diffraction grating acts as a Fourier Transformer by splitting up incident light into its constituent colours (in a similar manner to that of a prism, but with far higher resolution). You have probably observed this effect when viewing a CD, bands of colours which are due to optical diffraction of light from the closely spaced tracks on its surface as shown in Figure 7.12. The reason why the diffraction pattern repeats is born out by the Dirichlet kernel, when extended it actually appears as shown in Figure 7.13.

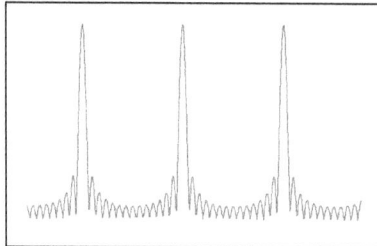

Figure 7.13: The repetitive character of the Dirichlet Kernel

The greater the number of lines per unit length on a diffraction grating the higher resolution. This corresponds to the number of side-lobes between each major peak in Figure 7.13. A large body of research has been performed and is currently going on into the field of processing optical signals which is leading towards future high speed computers based on optical technology.

What you have gained from this Chapter
1. How the use of the Z-transform and the shift theorem on difference equations.

2. An understanding of the concept of the zero.
3. How to use the Inverse Z-transform.
4. Passing from the Z-plain to the frequency domain.
5. An understanding of how sums and products work together.
6. An understanding of the unit circle and how zeros are located.
7. Regions of Convergence in the Z domain.
8. The effect of digitising a square wave and what it becomes in the frequency domain.
9. An insight into the role of the Dirichlet Kernel and how it extends to optical processing.

DADiSP skills you have acquired from this Chapter

1. *gimpulse* - creating an impulse function.
2. *delay* - adding a delay in a waveform
3. *W1 - W2* - subtracting the trance of one window from that of another.
4. *zfeq* - converting from the Z-pane to the frequency domain
5. *spectrum* - obtaining a spectrum of a waveform.
6. *zplane* - showing the positions of zeros on the unit circle.

DADiSP Extra

Very often when collecting measurement data, you will need to plot the data and find the *best curve fit* for the data. If the data points approximate to a straight line this is called *linear regression* which can be performed in DADiSP. First generate data values with a gradient and add random numbers. Next perform *linreg(w1,1)* in W2 and *overlay* the result on W1, the result after editing is shown in Figure 7.14.

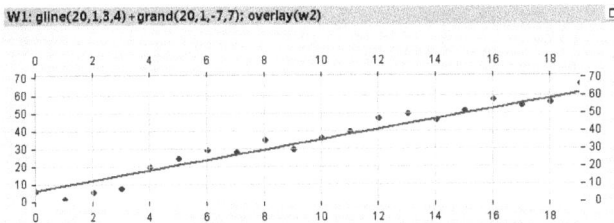

Figure 7.14: Linear regression with best line fit

You will also see the intercept and gradient values on the screen.

8. Finite Impulse Response (FIR) Filters

The first class of digital filters which receives considerable attention is the Finite Impulse Response filters or FIR filters. Over the pass few decades they have been used in countless applications and are fundamental to the science of digital signal processing. One of their more attractive features is that they are unconditionally stable - a serious requirement when using digital signal processing in critical systems.

8.1 FIR Filters from Convolution

The starting point for a discussion on FIR filters is the convolution process discussed in detail in Chapter 4. The convolution integral is given by,

$$y(t) = \int h(t') \, x(t - t') dt' \qquad \text{... 8.1}$$

A full discussion of this has been presented in Chapter 4 and it is recommended you read through that section before proceeding. Also a digitised version of the convolution was presented in section 5.3, the result was,

$$y(m) = \tfrac{1}{N} \sum_{n=0}^{n=N-1} h(n) \, x(m - n) \qquad \text{... 8.2}$$

With the substitution,

$$b_n = \tfrac{h(n)}{N} \qquad \text{... 8.3}$$

this leads directly to the general expression for a FIR filter which is,

$$y(m) = \sum_{n=0}^{n=N-1} b_n \, x(m - n) \qquad \text{... 8.4}$$

The characteristics of the filter are determined by the values of the coefficients $\{b\}$. When a FIR is designed, values are calculated for these coefficients and there are several design methods. One way of envisaging the operation of a FIR filter is to consider Figure 8.1 which shows a set of samples derived from a waveform. This filter can be expressed as,

$$y(m) = b_0 x(m) + b_1 x(m - 1) + b_2 x(m - 2) \qquad \text{... 8.5}$$

The filter takes three data points only and it weights (multiplies) each data point by a b coefficient.

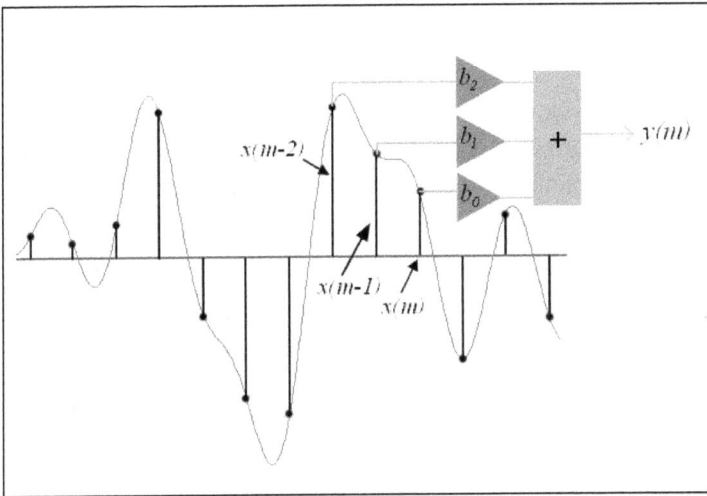

Figure 8.1: A method for visualising a FIR filter

Once the calculation in Eq:8.5 has been completed the output $y(m)$ is produced. The filter then moves to the right by one data sample and performs the calculation afresh. You can think of the FIR filter as *riding* the samples and producing an output steam of processed data $\{y\}$. Although in this example there are only three samples, in reality it's not unrealistic to have FIR filters containing 60 or 70 coefficients. When linked to an ADC, as each new sample emerges from it, the filter process operates on the new sample and the previous N-1 samples. The whole calculation must be completed before the next sample emerges from the ADC. This may be quite a challenge if the FIR filter has 60 coefficients. The generalised FIR filter takes on the form shown in Eq:8.4. When a Z-transform is performed on this expression we obtain,

$$Y(z) = X(z) \sum_{n=0}^{n=N-1} b_n z^{-n} \qquad \ldots 8.6$$

And the transfer function is

$$H(z) = \sum_{n=0}^{n=N-1} b_n z^{-n} = \prod_{n=1}^{n=N-1} \frac{(z-z_n)}{z^n} \qquad \ldots 8.7$$

The right hand side of Eq:8.7 is the product of all the zeros in the FIR filter and whenever $z = z_n$ the transfer function $H(z)$ will be zero. These zeros may be found anywhere in the unit circle and their influence will depend on how close they are to the perimeter of the unit circle.

8.2 Schematic of FIR Filters

In Chapter 5.5 we introduced the idea of a schematic to represent a filter, we can use this method to show how FIR filters are configured. Consider the third order FIR filter,

$$y(m) = b_0x(m) + b_1x(m-1) + b_2x(m-2) + b_3x(m-3) \qquad ... 8.8$$

You will observe from Figure 8.2 the three fundamental operations which are performed in a digital signal processor, multiplication, addition and data storage.

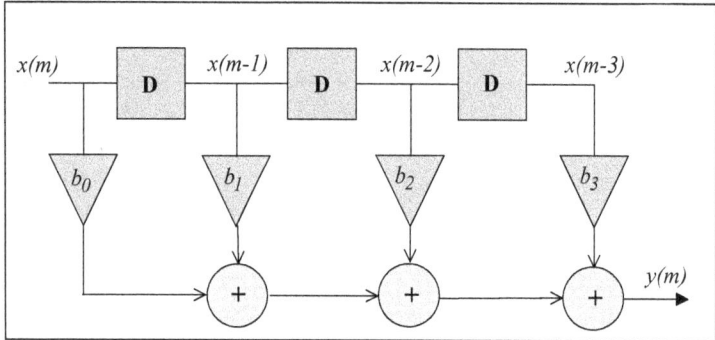

Figure 8.2: A schematic of the FIR filter in Eq:8.9

As previously stated, these processors have been specifically designed to maximise the execution rate of these operations. Although the filter in Eq:8.8 is limited to only three memory elements, there is no reason why it cannot be extended to a greater number. In a digital filter the combination of a memory element and a multiplier is often called a *tap*. Filters are therefore rated as 10-tap or 20-tap depending on their complexity. Also, the configuration shown in Figure 8.2 is called a *direct realisation*.

8.3 Filter Profiles

FIR filters usually fall into one of four groups and these are low pass (LP), high pass (HP), band pass (BP) and band stop (BS) which is sometimes known as a notch filter. Figure 8.3 shows these filter profiles.

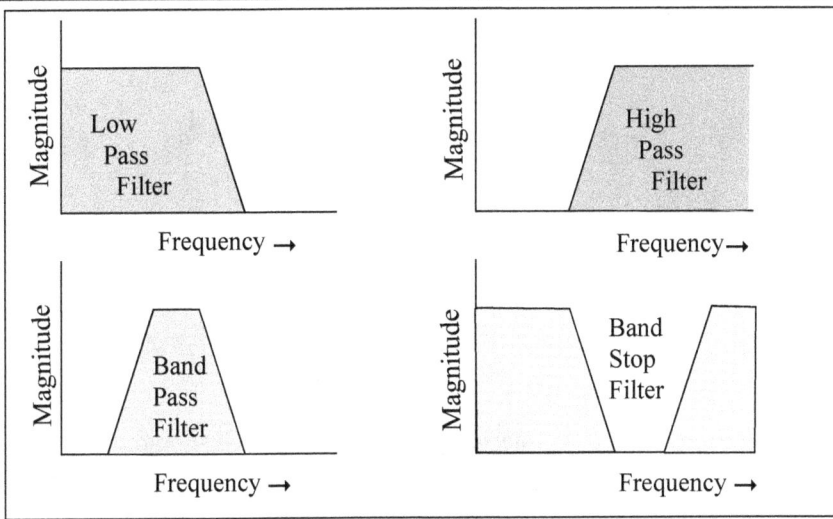

Figure 8.3: The four types of filter

The interesting fact relating to digital filters, the HP, BP and BS filter can all be derived from the LP filter. If $h(n)_{LP}$ is the impulse response of a low pass filter, then for

High pass filter

$$h(n)_{HP} = (-1)^n h(n)_{LP} \qquad \ldots 8.9$$

Band pass filter

$$h(n)_{BP} = 2\cos(n\omega_o\varDelta) h(n)_{LP} \qquad \ldots 8.10$$

Band stop filter

$$h(0)_{BS} = 1 - h(0)_{BP} \qquad \ldots 8.11$$

$$h(n)_{BS} = -h(n)_{BP}, n = \pm 1, \pm 2..$$

In the above expression ω_o is the central frequency of the BP filter. You can think of a HP filter as a reflexion of a LP filter.

There is a command in DADiSP which generates a FIR filter and the following is an example of how it's used. First create a signal which has a low frequency (5 Hz) and a high frequency (40 Hz) as shown in Figure 8.4.

W1: gsin(300,.01,5)+gsin(300,.01,40)

97

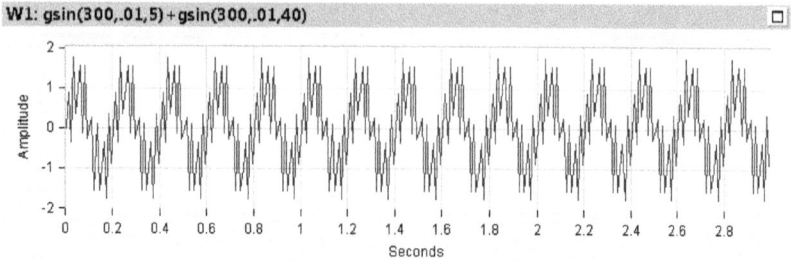

W1: gsin(300,.01,5)+gsin(300,.01,40)

Figure 8.4: A waveform containing a low frequency
and a high frequency

By performing high pass filtering, the lower frequency should be removed.

W2: Filteq({-.1,.2,-.4,.6,-.8,1,-.8,.6,-.4,.2,-.1 }, {1},w1)

This filter contains eleven coefficients and will operate on W1 and the results can be seen in Figure 8.5.

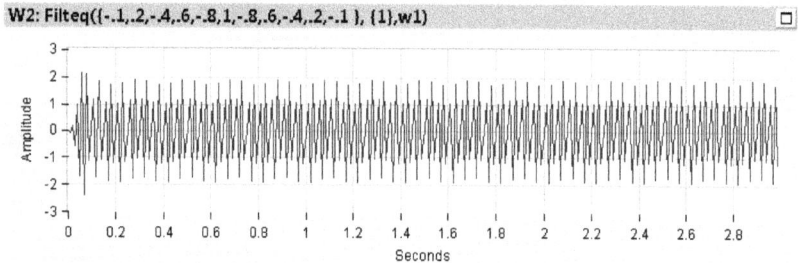

W2: Filteq({-.1,.2,-.4,.6,-.8,1,-.8,.6,-.4,.2,-.1 }, {1},w1)

Figure 8.5: The filtered version of Figure 8.4

The spectrum for the input waveform is shown in Figure 8.6.

W3: Spectrum(w1)

Now compare this with the spectrum of output of the filter as shown in Figure 8.7.

W4: Spectrum(w2)

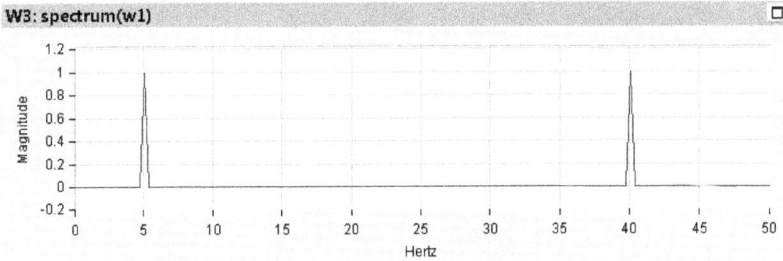

Figure 8.6: The spectrum of the input waveform (Figure 8.4)

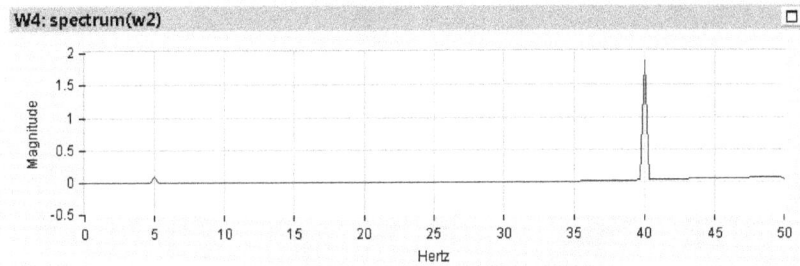

Figure 8.7: Spectrum of the filtered waveform

You will observe, in Figure 8.7 the low frequency component has almost been filtered out leaving only the high frequency (40Hz) component. To obtain a complete transfer function of the filter, the input waveform should be an impulse.

W1: gimpulse(300,.01,0)

When this is created in W1, the true impulse response will appear in W2 which is shown in Figure 8.8. It is worth noticing the duration of the impulse response (0.1s) in Figure 8.8 and its transfer function is shown in Figure 8.9. You will observe from Figure 8.9 the high pass profile of the filter. Although the roll-off of the filter is by no means steep, it does demonstrate the use of the Filteq command in DADiSP for realising digital filter profiles which will be explored further later in this chapter.

W2: Filteq({-.1,.2,-.4,.6,-.8,1,-.8,.6,-.4,.2,-.1 }, {1},w1)

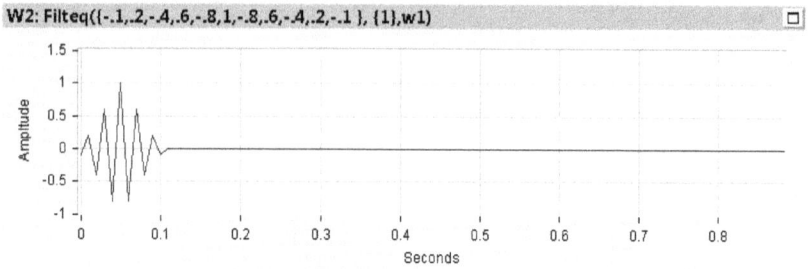

Figure 8.8: The impulse response of the high pass filter

W4: spectrum(w2)

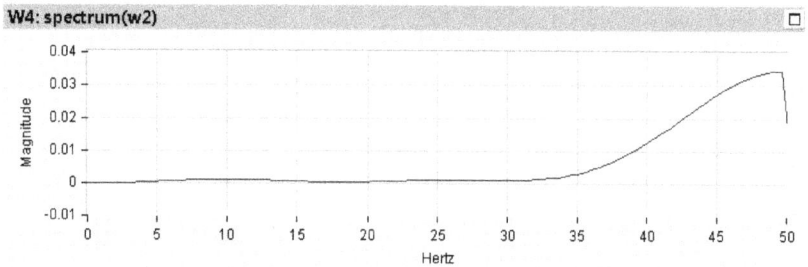

Figure 8.9: Transfer function of the high pass filter

8.4 Sinc Functions and FIR Filters

In Chapter 6.2 it was shown that a rectangular pulse in the time domain produces a sinc function in the frequency domain. It will come as no surprise that a sinc function in the time domain will produce a rectangular function in the frequency domain as illustrated in Figure 8.10.

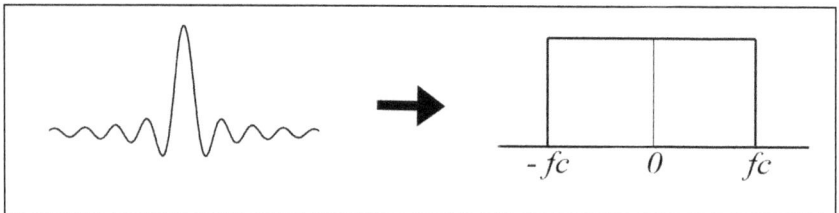

Figure 8.10: A sinc function in the time domain transforms into a rectangular pulse in the frequency domain.

In Figure **8.10** you will observe a rectangular profile in the frequency domain, if you consider only the right half (from *0* to *fc*) it takes on the appearance of a low pass filter. This implies that if you design a FIR filter whose impulse response is a sinc function it will behave as a low pass filter. In Figure **8.10** *fc* is referred to as the *cut-off frequency* and this imaginative filter is called a *brick wall filter* for obvious reasons, which incidentally cannot be realised. In this ideal filter every frequency beyond *fc* is stopped from passing through. The value of *fc* is determined by the shape of the sinc function. This was discussed in some detail in Chapter 6. If the sinc function is very narrow, *fc* is large, on the other hand if the sinc function is broad, the value of *fc* is low. Therefore what occurs is the inverse effect as you move from the time domain to the frequency domain and vice versa. This can be illustrated by analysing the effect in DADiSP. First create a narrow sinc function as shown in Figure 8.11.

W1: gsinc(300,.01,100,-100)

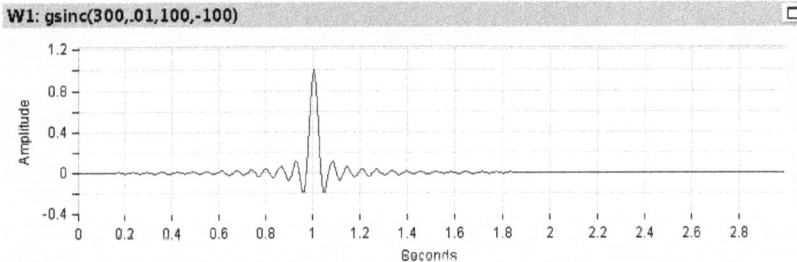

Figure 8.11: A narrow sinc function which arrives 1 second after its start

Now observe the spectrum of the sinc function as shown in Figure 8.12.

W3: spectrum(w1)

You will observe the spectrum in Figure 8.12 has an appearance very similar to a brick-wall filter shown in Figure 8.10 where the cut-off frequency is just over 15Hz. In effect the cut-off frequency *fc* is determined by the width of the sinc function. You will also notice some slight ripple in the pass band region of the filter and this artefact will be discussed later.

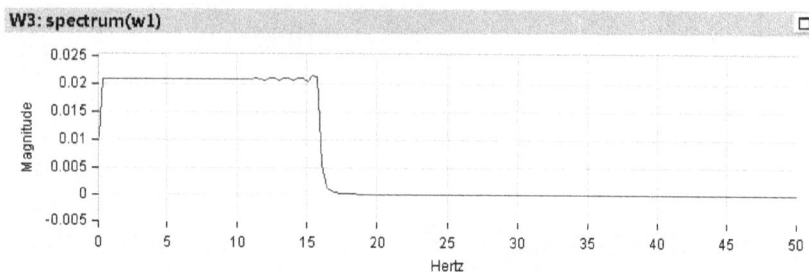

Figure 8.12: Spectrum of narrow sinc function

Now consider the case of a broad sinc function as shown in Figure 8.13.

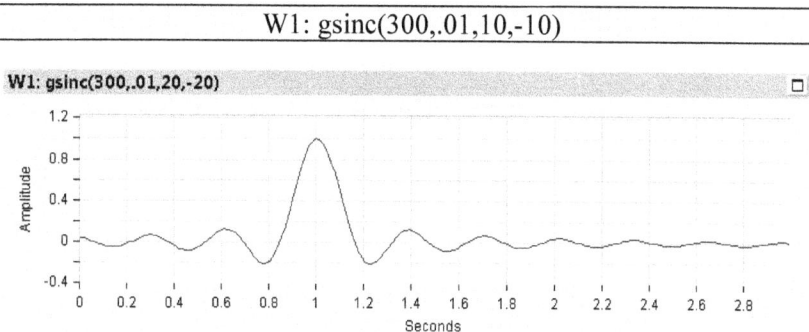

W1: gsinc(300,.01,10,-10)

Figure 8.13: A broad sinc function

The Spectrum of the sinc function shown in Figure 8.13 is shown in Figure 8.14.

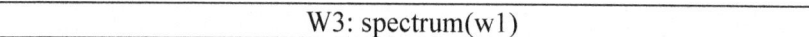

W3: spectrum(w1)

As can be observed from Figure 8.14 the width of the band pass region has been greatly reduced as a consequence of broadening the sinc response function. In general, the wider the sinc function, the narrower the transfer function. The reverse is also true, the narrower the sinc function, the broader the transfer function.

102

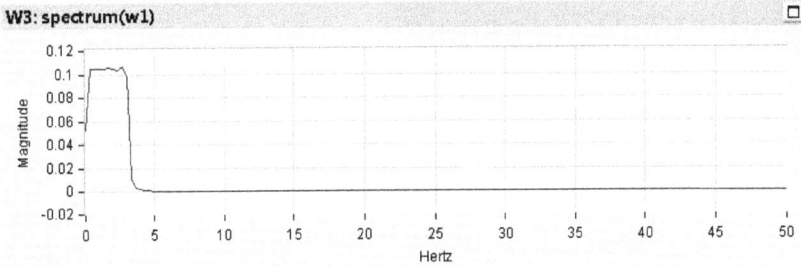

Figure 8.14: The spectrum of a broad sinc function

8.5 Truncating the Sinc Function

The sinc function has an infinite duration and in reality this cannot be realised and there are definite effects when the duration of sinc function is truncated in the time domain. In the first instance, it gives rise to ripple in both the pass band and the stop band - this is known as *Gibbs Phenomena*. It also causes a finite roll-off on the falling edge of the filter as shown in Figure 8.15.

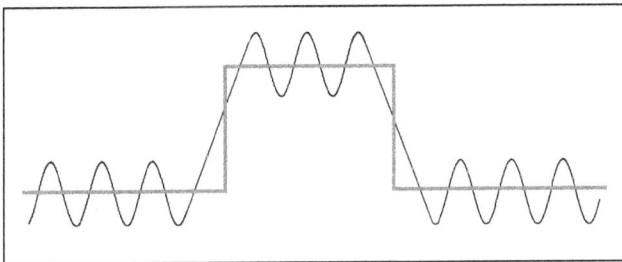

Figure 8.15: Effect if truncating the sinc function

It's the profile in Figure 8.15 which makes the design of FIR filters interesting. A section of the low pass filter's transfer function is used to determine the characteristics as shown in Figure 8.16. The design parameters, include,

- pass band ripple
- stop band ripple
- start of transition frequency
- end of transition frequency
- Sampling frequency.

These parameters are derived from a transfer function similar to that in Figure 8.16 and this will be discussed in Chapter 9.

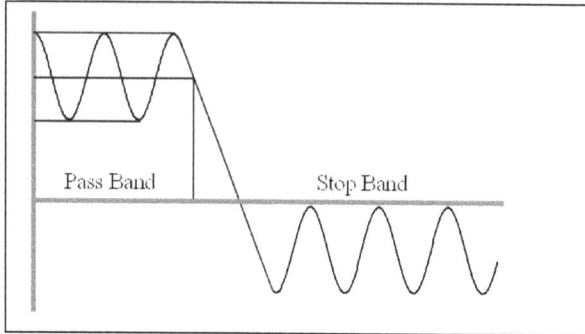

Figure 8.16: The regions of the low pass filter are identified

8.6 Digitising the Sinc Function

When looking at Eq:8.7 it contains a set of coefficients $\{b\}$ and an interesting question - where do these coefficients come? An important type of FIR filter arises where the impulse response is *symmetrical* as in the sinc function. Such a filter has *linear phase* and the significance of this will be discussed shortly. It is particularly important in audio applications. The coefficients $\{b\}$ come from digitising the sinc function and it is these numbers which are the filter's coefficients. Figure 8.17 shows a digitised sinc function. You will observe from the digitised values the left had side is the same as the right hand side - hence they are *symmetrical*. Only half the coefficients are unique so you'll find the coefficients similar to that in the table on the left. Very often the filters coefficients are far more numerous - sometime up to 70 of them. Also the precision is much greater. The coefficients given in the table on the left have only 4 digits, typically they would have a precision of 9 or 10 digits. Reducing the precision of the coefficients is referred to as *quantisation*.

$$H(01) = -0.1361E{-}01 = H(27)$$
$$H(02) = 0.3479E{-}02 = H(26)$$
$$H(03) = 0.1114E{-}01 = H(25)$$
$$H(04) = 0.1666E{-}01 = H(24)$$
$$H(05) = 0.1280E{-}01 = H(23)$$
$$H(06) = -0.3320E{-}02 = H(22)$$
$$H(07) = -0.2616E{-}01 = H(21)$$
$$H(08) = -0.4207E{-}01 = H(20)$$
$$H(09) = -0.3476E{-}01 = H(19)$$
$$H(10) = 0.5533E{-}02 = H(18)$$
$$H(11) = 0.7507E{-}01 = H(17)$$
$$H(12) = 0.1552E{-}00 = H(16)$$
$$H(13) = 0.2193E{-}00 = H(15)$$
$$H(14) = 0.2437E{-}00 = H(14)$$

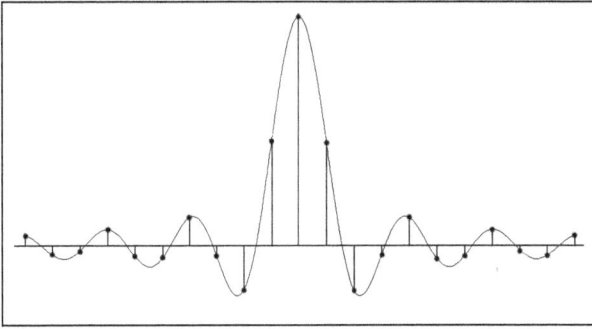

Figure 8.17: A digitised sinc function

8.7 FIR Filters and Unit Circle

It has been shown that in the Z-plane FIR filters have zeros and it is of interest to relate the position of these zeros to the performance of a filter. To illustrate this concept an example will be considered. Given a FIR difference equation,

$$y(m) = x(m) - \sqrt{2}\,x(m-1) + x(m-2) \qquad \text{... 8.12}$$

Performing a Z-transform on this difference equation, we obtain the transfer function which is,

$$H(z) = \frac{z^2 - \sqrt{2}\,z + 1}{z^2}$$

$$= \frac{\left(z + \frac{1}{\sqrt{2}}(1+j)\right)\left(z + \frac{1}{\sqrt{2}}(1-j)\right)}{z^2} \qquad \text{... 8.13}$$

First create an impulse to feed into the FIR filter.

W1: gimpulse(300,.01,0)

Now create the filter for Eq:8.13.

W2: Filteq({1, -1.414, 1}, {1}, w1)

Create the spectrum of the output of the filter in W3 as shown in Figure 8.18.

W3: 20*log(spectrum(w2))+100

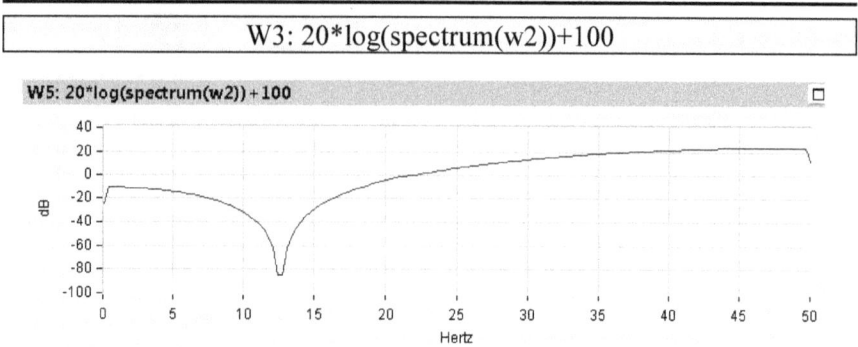

Figure 8.18: Spectrum of FIR filter with two zeros

Figure 8.18 shows the spectral position of the zero. Now examine this in the Z-plane which is shown in Figure 8.19.

W4: zplane({1, -1.414, 1}, {1})

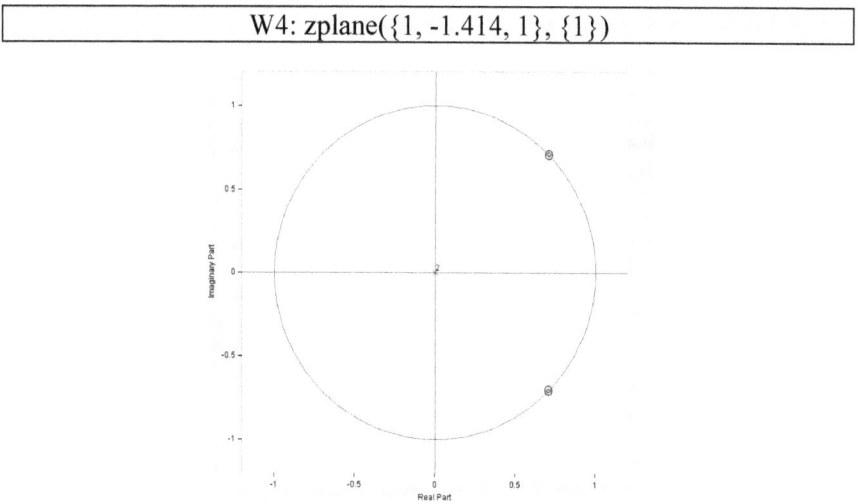

Figure 8.19: Unit circle showing the locations
of the zeros for the filter in Eq:8.12

To verify the value of the zero position, convert the first root into its radial coordinates: $|z1| = 1$, $\theta = 45^\circ$. From Figure 8.18, $180^\circ \equiv 50$ Hz, therefore 45° is equivalent to,

$$\frac{50}{180} \times 45 = 12.5 \text{ Hz}$$

which corresponds to the zero position in Figure 8.18. From this example you are able to appreciate how the zero positions occur in the unit circle. Consider the FIR filter with a transfer function as shown Figure 8.20,

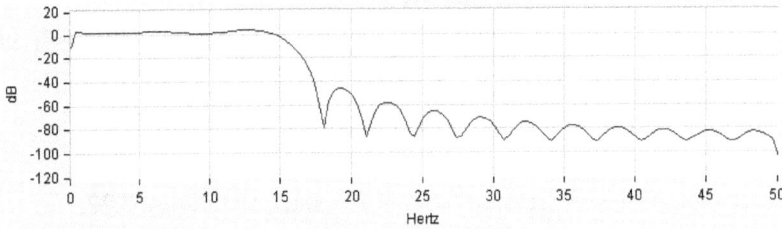

Figure 8.20: FIR filter with several coefficients

The distribution of zeros in the Z-plane for this filter is shown in Figure 8.21.

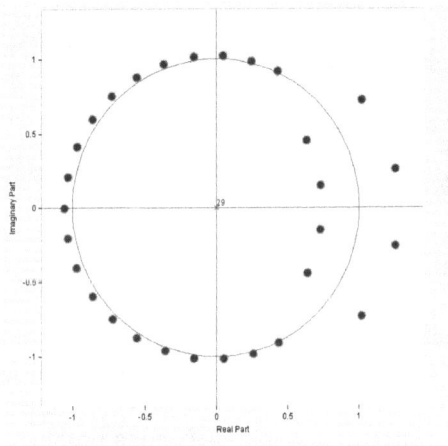

Figure 8.21: The zero positions for the filter shown in Figure 8.20
(the zeros are shown as large dots to make them more visible)

Figure 8.20 is a low pass filter and you will observe from Figure 8.21 as you follow the perimeter from 0Hz, in either direction (the band pass region), you reach the first zero on the perimeter. This is the first zero after the roll-off edge. Thereafter its the stop band region is in effect a sequence of *closely spaced zeros*. The number of zeros corresponds to the number of coefficients in the filter. The attenuation in the stop band is increased by having even more tightly spaced zeros.

8.8 Cascade Realisation of FIR Filters

In Eq:8.7 a FIR was represented by first order factorising, alternatively a FIR filter may be factorised as a product of second order sections,

$$H(z) = \prod_{n=1}^{n=N/2} [b_{n0} + b_{n1}z^{-1} + b_{n2}z^{-2}] \qquad \dots 8.14$$

A schematic of this expression for $N = 4$ is shown in Figure 8.22,

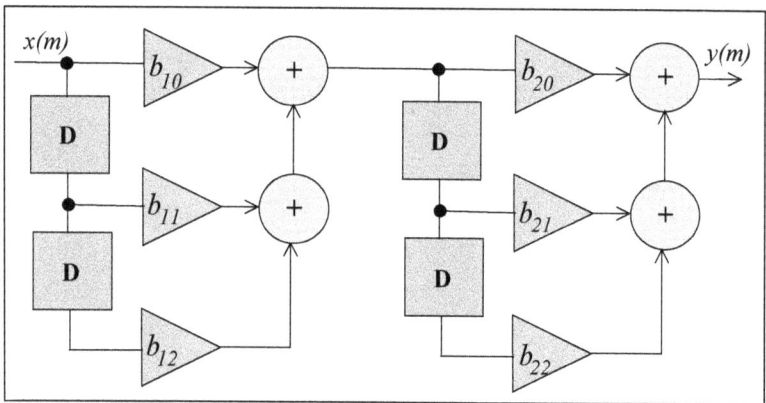

Figure 8.22: A schematic for a fourth order FIR filter

Figure 8.22 is called a *cascade realisation* as it comprises second order sections where the processed signal propagates from left to right through each section. These are useful when considering implementation on digital signal processors when scaling may become a problem. When multiplying two fractions the result is a smaller fraction. However when adding two fractions the result is a larger fraction. It is sometimes advantageous to *interlace* the sequence of multiplications and additions in order to maximise numerical precision. For example, consider a precision of 4 decimal places,

.	2	0	0	0

Multiply by 0.2,

.	0	4	0	0

Multiply by 0.2,

Multiply by 0.2

| . | 0 | 0 | 8 | 0 |

Multiply by 0.2

| . | 0 | 0 | 1 | 6 |

Multiply by 0.2

| . | 0 | 0 | 0 | 3 |

| . | 0 | 0 | 0 | 0 |

You will observe the number has fallen off the end and this is a major problem you have to guard against when programming DSPs. You would therefore not have a sequence of multiplications or additions; you would interlace them to ensure there is no overflow or a product falling out of a register.

8.9 Quantising Filter Coefficients

Coefficients are used in the implementation of FIR filters on Digital Signal Processors. Every processor has limited precision. The coefficients therefore have to be reduced in precision for them to be accommodated on the processor. This is called quantisation and this can have a serious effect on the expected performance of the filter. When programming a DSP, the assembly language normally deals with hexadecimal numbers (numbers with a base of 16). The coefficients are therefore conditioned to *fit* onto a DSP. Let's consider an example, the coefficient value is,

$$0.8695326$$

and we condition this for a 24-bit processor. The maximum value of a 24 bit number is

$$2^{23} = 8,388,608 = \$7F\ FFFF.$$

As the precision of the sample number is reduced (by replacing the least digit by a 0), you are able to see the effect it has on the final Hex number which would be implemented on the processor,

$$0.\mathbf{8695326} \times 8,388,608 = 7,294,168 = \$6F\ 4CD8$$
$$0.\mathbf{8695320} \times 8,388,608 = 7,294,163 = \$6F\ 4CD3$$
$$0.\mathbf{8695300} \times 8,388,608 = 7,294,146 = \$6F\ 4CC2$$
$$0.\mathbf{8695000} \times 8,388,608 = 7,293,894 = \$6F\ 4BC6$$
$$0.\mathbf{8690000} \times 8,388,608 = 7,289,700 = \$6F\ 3B64$$
$$0.\mathbf{8600000} \times 8,388,608 = 7,214,202 = \$6E\ 147A$$

Where the $ sign indicates a Hex number. The most significant bit is given over to the sign (0 for positive and 1 for negative). You will observe in the above calculations as the precision is reduced it makes a serious difference in the final Hex value. You would be right in thinking this has a serious effect of the operation of any algorithm. The quantisation of numbers requires attention and one of the reasons why most DSPs today have precision's of 24-bits or greater is to minimise this effect. When the coefficients are subjected to quantisation the effects of round-off must be taken into consideration. Figure 8.23 shows the effect of quantising the coefficients down to 8-bit has on the performance of a FIR filter.

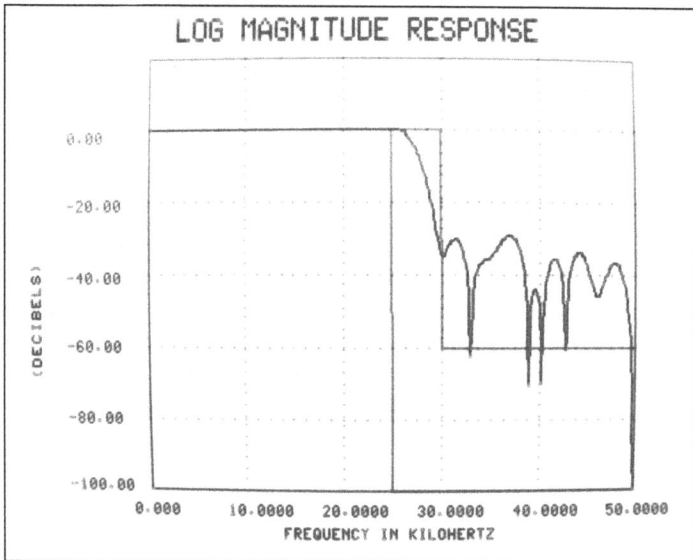

Figure 8.23: Effect of quantising the FIR coefficients to 8-bits

You will observe in Figure 8.23 the required filter envelope with a start transition at 25kHz, an end of transition at 30kHz and a required attenuation at -60dB. You will also observe the actual transfer function which strays well outside the specification envelope. Although in practise you would not quantise down to 8-bits, nevertheless you can appreciate that quantising coefficients can cause the filter transfer function to stray well beyond its intended specification.

What you gained from this Chapter

1. An understanding how FIR filters are derived from the concept of convolution.
2. How to construct a schematic of a FIR filter.
3. How to define filter profiles.
4. Understanding how the sinc function and FIR filters are related.
5. The effect of truncating the sinc functions
6. Digitising the sinc function
7. Relating the performance of FIR filters to the unit circle.
8. How to realise a cascade of FIR filter elements
9. The effect of quantisation of FIR filter coefficients

The DADiSP skills you have acquired from this Chapter

1. *gsin-* to generate a sine wave
2. *filteq* - to create a filter from a set of coefficients
3. *spectrum* - to obtain a spectrum of a signal
4. *gimpulse* - to generate an impulse function
5. *gsinc* - to generate a sinc function
6. *20*log10(spectrum)* - to obtain a spectrum with a dB magnitude
7. *zplane* - to create a unit circle showing the position of zeros

---

DADiSP Extra

Polynomials are often of great interest and below is an example of 5th order polynomial with five real roots. Click on $f_x \rightarrow$ Generate Data $\rightarrow$ Y=F(X). When the dialogue window opens, in the Y=F(X) field enter, $(x+3)*(x+2)*(x-1)*(x-2)*(x-4)$. Select a range for $x \in [-3.1, 4.2]$ with an increment of 0.01. The result is as shown in Figure 8.24. Notice how the trace passes through the *zero axis* five times.

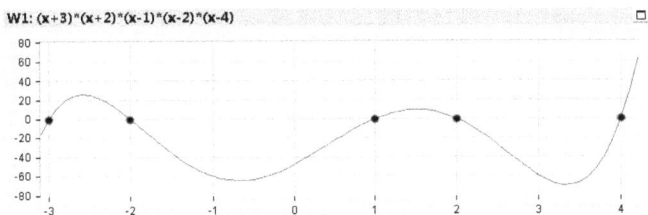

W1: (x+3)^(x+2)^(x-1)^(x-2)^(x-4)

Figure 8.24: Plot of a 5th order polynomial

111

9. FIR Filter Design

Having looked at several features of FIR filters, it now remains to discuss the various methods of designing them. Attention has been focused on the importance of the sinc function and the fact the design coefficients are derived from digitising it. One requirement, especially for real-time applications, is for the filters to execute in minimum time - that is to achieve the required specification and execute the filter algorithm with a minimum of calculations.

9.1 Window Design

In Chapter 6 the sinc function was considered in detail and the effect of truncating its profile was discussed in Chapter 8. Instead of chopping the sinc profile so abruptly an alternative method involves the use a shaping profile, known as a *data window*. As previously stated the skirts of the sinc function approach $|\infty|$ which is somewhat impracticable. By imposing a data window on the sinc profile its skits reach zero very quickly. The window therefore discriminates against the ripples which are well away from the central lobe. There are several of these data windows available and they are also used in spectral analysis. We shall work with a number of these windows which can be found in DADiSP. First create a sinc function as shown in Figure 9.1.

W1: gsinc(40,.01,100,-15)

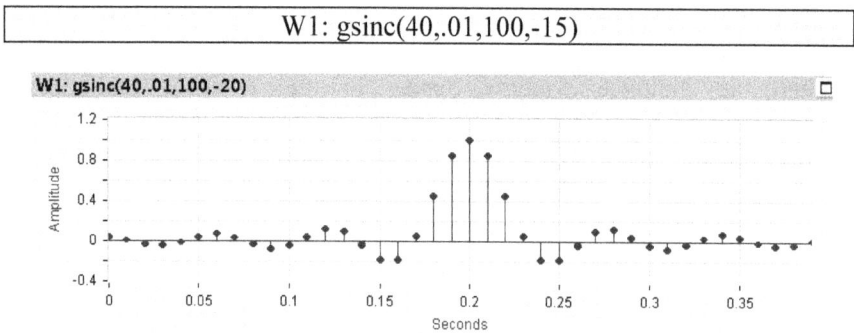

Figure 9.1: A sinc function reaching a peak at 0.2s

Next generate an impulse function to pass through the sinc filter.

W2: gimpulse(100, .01,)

Once the impulse function has been generated, we can convolve the sinc function with the impulse function to give the result as shown in Figure 9.2.

W3: conv(w3,w4)

W3: conv(w1,w2) □

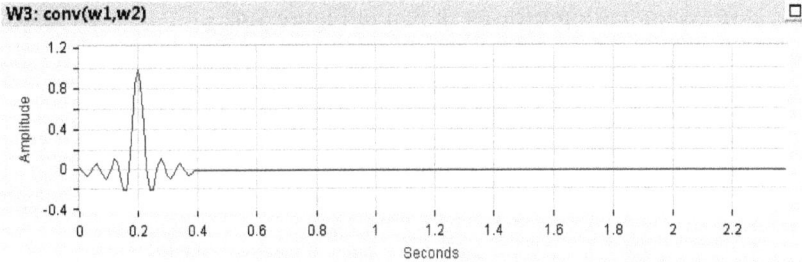

Figure 9.2: Impulse response of the sinc function

The spectrum of the impulse response in Figure 9.2 is shown in Figure 9.3. 80dB has been added to scale the spectrum to zero.

W4: 20*log(spectrum(w5))

W4: 20*log10(spectrum(w3))+30 □

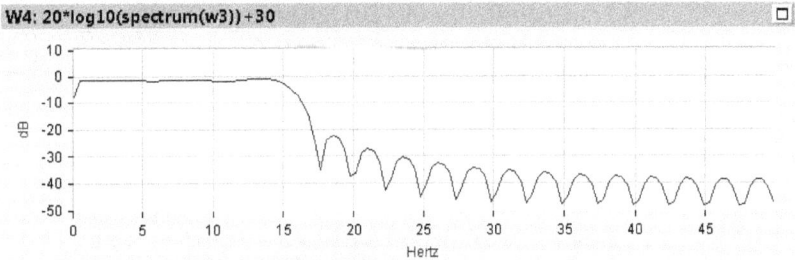

Figure 9.3: The spectrum of the impulse response in Figure 9.2

This is the characteristic spectral profile of a low pass filter with ripple in the stop band (beyond 17 Hz). The minimum attenuation is -22dB (confirm using right click, **Cursor** → **Crosshair**) and the purpose of applying a

---

data window is to reduce the ripple in the stop band thereby increasing the attenuation.

Hanning Window

In the first instance we are going to use the *Hanning Window*. This window has the form,

$$w_{Hann}(n) = 0.5[1 - \cos(\tfrac{2\pi n}{N-1})] \qquad \text{...9.1}$$

This expression has a profile as shown in Figure 9.4.

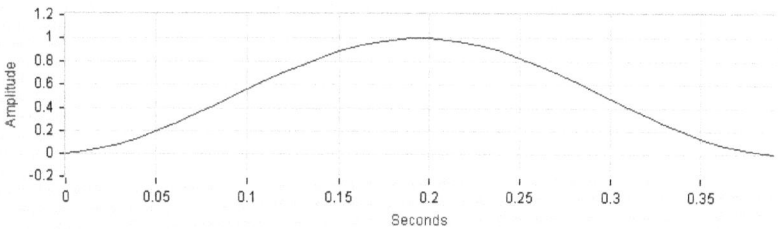

Figure 9.4: The profile of a Hanning Window

You will observe from this profile the peripheral values are zero, when imposed on a sinc function it effectively reduces its peripheral values to zero. A result of a windowed sinc function is shown in Figure 9.5.

W1: gsinc(40,.01,100,-20)*ghanning(40,.01)

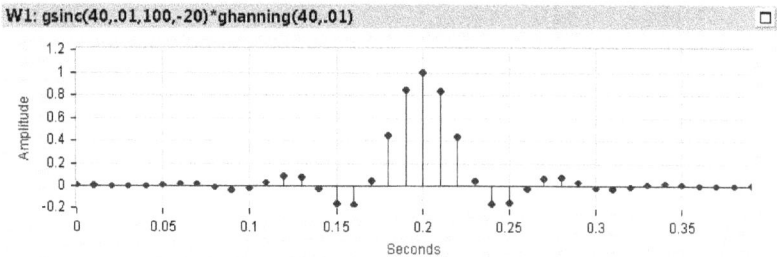

Figure 9.5: A Hanning Window applied to the sinc function

You will observe from comparing Figure 9.1 with Figure 9.5 a reduction in many of the coefficient values. You will also observe the ripple has been suppressed up to 0.05 seconds and beyond 0.35 seconds. The transfer function of the new sinc profile is shown in Figure 9.6 along with the transfer function of the sinc function without the windowing.

W4: 20*log10(spectrum(w3))+30; overlay(w5)   □ x

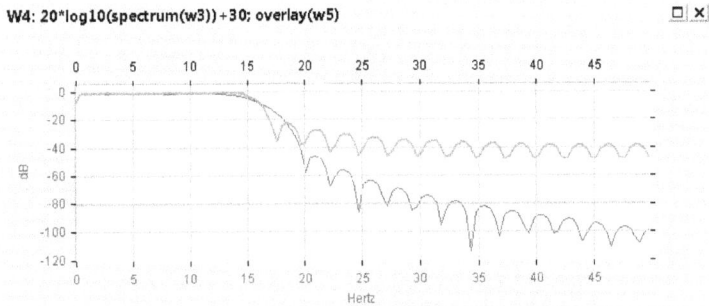

Figure 9.6: Transfer function with and without the Hanning Window

The ripple in the stop band has now been reduced to -45dB (verify this for yourself; right click and select **Cursor→Crosshair**). You will also observe a shift in the roll-off of the Hanning Windowed transfer function. The first zero for un-windowed transfer function occurs at 17.5Hz whereas for the windowed data it is 20 Hz. There are other windows from which to choose and the next application will be that of the Blackman Window.

Blackman Window
This window has the form,

$$w_{Blackman}(n) = 0.42 - 0.5\cos(\tfrac{2\pi n}{N-1}) + 0.08\cos(\tfrac{4\pi n}{N-1}) \qquad ...9.2$$

Imposing this window on the sinc function can be achieved by the command,

W1: gsinc(40,.01,100,-20)*gblackman(40,.01)

The effect on the transfer function is shown in Figure 9.7. The ripple is now -75dB down, but this comes at a cost, the pass band region is broadened. When compared with Figure 9.3 where the first zero occurs at 17.5Hz. The first zero has now moved to 25Hz. Using data windows certainly suppresses the side-lobes on FIR filters - it is very much *swings*

115

and roundabouts, where there is a gain in one feature you loose in another. Another well know window used in FIR filter design is the Kaiser-Bessel.

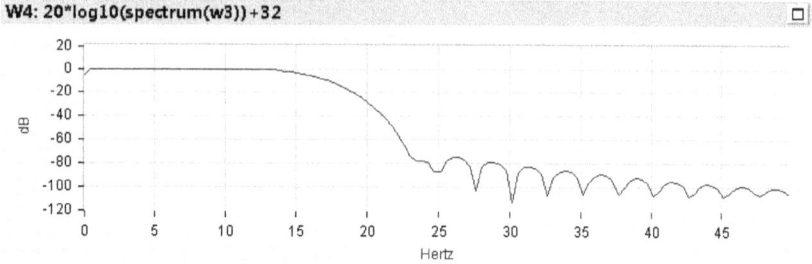

Figure 9.7: The effect of imposing a Blackman Window on a sinc function

Kaiser-Bessel

All data windows have a similar profile of discriminating against the data values on the peripheral of the data set. The Kaiser-Bessel window takes the form

$$w_{Kaiser}(n) = \frac{I_o\left[\beta\sqrt{1-\left(\frac{2n}{N-1}\right)^2}\right]}{I_o(\beta)} \qquad \dots 9.3$$

where I_o is the *zero order modified Bessel function of the first kind* and β is the *shape factor*. Applying this window to a sinc function requires the command,

W1: gsinc(40,.01,100,-20)*gkaiser(40,.01,10)

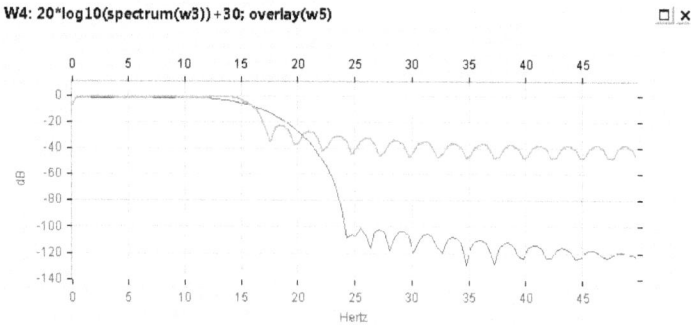

Figure 9.8: Transfer function of sinc function with and without the Keiser-Bessel Window

The resulting transfer function is shown in Figure 9.8 where the shape factor β has been allocated the value of 10. The ripple is now down to -99dB, but the first zero occurs at 26Hz - a significant broadening. You can also see in Figure 9.8 the transfer function for the unwindowed (raw) sinc function. It can be appreciated the Kaiser-Bessle Window is very effective at suppressing the ripple in the stop band. Using data windows is a viable method of designing FIR filters and is found in many commercial software packages for FIR design. As stated, the more you suppress the ripple in the stop band the greater the broadening of the pass band region.

Amplitude Modulation

The effect of broadening in the transfer function arise from the fact that an applied data window is in fact an *amplitude modulation*.

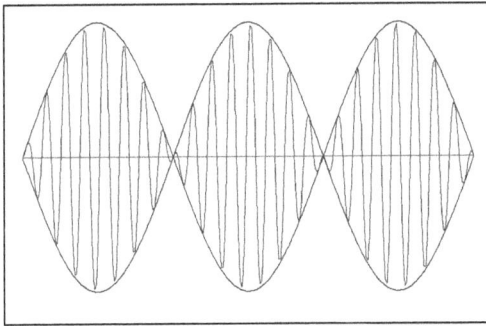

Figure 9.9: An example of amplitude modulation

Amplitude modulation is achieved by multiplying two waveforms together. Using the identity,

$$\sin(A) + \sin(B) = 2\sin(\tfrac{A+B}{2})\cos(\tfrac{A-B}{2}) \qquad \text{...9.4}$$

which may be written as

$$2\sin(f_1 t)\cos(f_2 t) = \sin(f_1 + f_2)t + \sin(f_1 - f_2)t \qquad \text{...9.5}$$

When considering the two waveform in Figure 9.9, the low frequency envelope corresponds to $f_1 - f_2$. The high frequency contained within the low frequency envelope corresponds to $f_1 + f_2$. Now consider only half a cycle as shown in Figure 9.10. This may not appear too significant, but compare it with Figure 9.11 which is a Hanning Window applied to a sine wave.

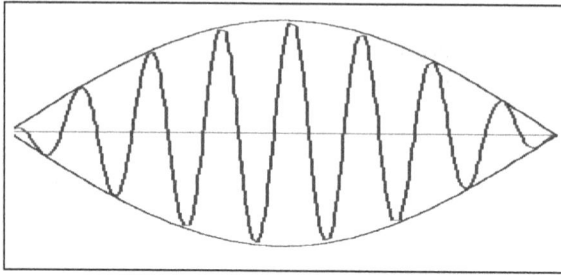

Figure 9.10: Half a cycle of a modulation

W5: gsin(500,.01,6)*ghanning(500,.01)

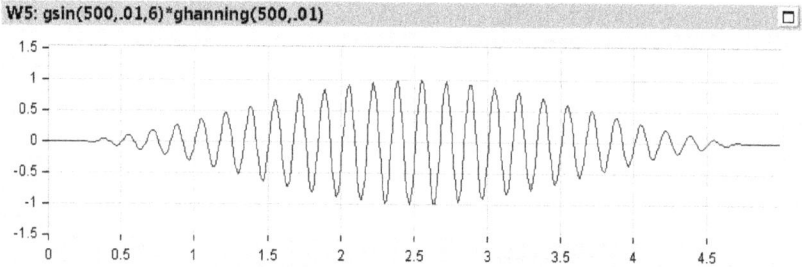

Figure 9.11: A Hanning Window applied to a sine wave

There is a strong resemblance between Figures 9.10 and 9.11. To resolve a frequency at least one cycle is required. The effect of only having half a cycle means the spectral profile gets *broadened* which we have observed with the application of a window, but it is only half a wave modulation. This is what gives rise to the broadening effect that is experienced when a window is applied to a data set.

Design Specifications
When using commercial software to design a FIR filter you will be required to enter certain specifications into the program. These are (refer to Figure 9.12): sampling frequency fs, pass band ripple $\delta1$, stop band ripple $\delta2$ (usually given in dB), start of roll-off transition frequency $f1$ and end of transition frequency $f2$. For band pass and band stop filters you will be asked to input additional specifications. After some practise you will find the design of FIR quite straight forward.

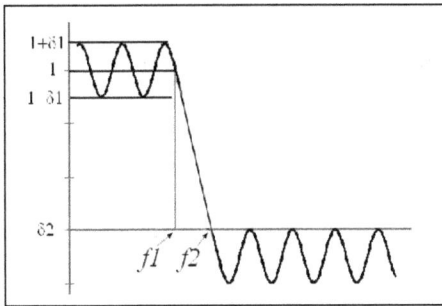

Figure 9.12: Specification needed for FIR filter design

If you require a band stop or a band pass filter you need to provide additional transition frequencies.

9.2 Equiripple Design

This method of FIR filter design was pioneered by the American engineers *Thomas Parks* and *James McClellan* in 1972. The objective of the *Equiripple Design* is to produce a set of filter coefficients which satisfy the specifications with a *minimum number of taps*. This enables the FIR filter to be executed in real-time with a minimum number of calculations. This is the objective of real-time algorithms - minimum execution time. Generally they will also require minimum code and memory space. The optimum FIR filter design is known as the *Parks-McClellan Algorithm*, it is a rather involved method for deriving the set of FIR filter coefficients from a set of specifications. Referring to Figure 8.12, you will observe ripple in both pass band and stop band. Whenever you have a *wavy* function or a ripple profile it can usually be represented (modelled) by a polynomial. The general expression for a polynomial is,

$$y(x) = \sum_{n=0}^{n=N} a_n x^n = \prod_{n=1}^{n=N} (x - x_n) \qquad ...9.6$$

where $\{a\}$ are the coefficients of the polynomial and $\{x_n\}$ are its roots which may be real or complex. In order to fit a polynomial to a function which has a minimum number of terms you require the use of the *Remez Exchange Algorithm*. This is a useful device for finding coefficients in a series. For example it's possible to find the approximate value of $\sin(x)$ from a series with a *finite* number of coefficients,

$$f(x) = a_0 + a_1x + a_2x^2 + a_3x^3 + a_4x^4 + a_5x^5 \simeq \sin(x) \qquad ...9.7$$

The Remez Exchange Algorithm enables you to determine the coefficients in Eq:9.7 to give the minimum value of $|f(x) - \sin(x)|$. Using these ideas Parks and McClellan where able to devise a very elegant method of calculating the coefficients in a FIR filter. Fortunately as a user you never have to use their algorithm numerically as its already been coded in FORTRAN and the majority of commercial software use it for filter design. Incidentally, a similar expression to Eq:9.7 is used to calculate the approximate values of many trigonometric and other functions.

9.3 Symmetrical FIR Filters

Considerable attention has been paid to the sinc profile and this is an example of a *symmetrical filter*. They have special properties which are particularly important to audio applications, in particular the phase characteristics. Consider the general FIR filter whose transfer function is,

$$H(\omega) = \sum_{n=0}^{n=6} h(n)e^{-jn\omega\Delta} \qquad ...9.8$$

Expanding this expression,

$$H(\omega) = h(0) + h(1)e^{-j\omega\Delta} + h(2)e^{-j2\omega\Delta} + h(3)e^{-j3\omega\Delta} +$$
$$h(4)e^{-j4\omega\Delta} + h(5)e^{-j5\omega\Delta} + h(6)e^{-j6\omega\Delta} \qquad ...9.9$$

This can be expressed as,

$$H(\omega) = e^{-j3\omega\Delta}[h(0)e^{j3\omega\Delta} + h(1)e^{j2\omega\Delta} + h(2)e^{j\omega\Delta} +$$
$$h(3) + h(4)e^{-j\omega\Delta} + h(5)e^{-j2\omega\Delta} + h(6)e^{-j3\omega\Delta}] \qquad ...9.10$$

Owing to the symmetry of the sinc function, $h(0) = h(6)$, $h(1) = h(5)$ and $h(2) = h(4)$, we can now pair up terms in this expression,

$$H(\omega) = e^{-j3\omega\Delta}\{h(0)(e^{j3\omega\Delta} + e^{-j3\omega\Delta}) + h(1)(e^{j2\omega\Delta} + e^{-j2\omega\Delta}) +$$
$$h(2)(e^{j\omega\Delta} + e^{-j\omega\Delta}) + h(3)\} \qquad ...9.11$$

This can be expressed as,

$$H(\omega) = e^{-j3\omega\Delta}\{2h(0)\cos(3\omega\Delta) + 2h(1)\cos(2\omega\Delta)$$
$$+ 2h(2)\cos(\omega\Delta) + h(3)\} \qquad ...9.12$$

If we let $a_0 = h(3)$ and $a_n = 2h(3-n)$, then

$$H(\omega) = e^{-j3\omega\Delta} \sum_{n=0}^{n=3} a_n \cos(n\omega\Delta) = |H(\omega)|e^{j\phi(\omega)} \qquad \dots 9.13$$

where the phase spectrum is,

$$\phi(\omega) = -3\omega\Delta \qquad \dots 9.14$$

The gradient of the line is -3Δ and the greater the number of taps in the filter the greater the gradient. A symmetrical response function can be expressed as,

$$H(\omega) = |H(\omega)|e^{j\phi(\omega)} \qquad \dots 9.15$$

The significance of Eq:9.15 is the fact all the real components are in $|H(\omega)|$ and the imaginary component is in the argument attached to the exponent. Looking at Eq: 9.14, the phase spectrum $\phi(\omega)$ is linear as shown in Figure 9.13 (Δ is a constant - the sampling interval).

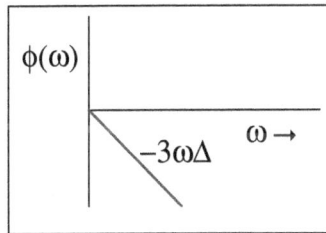

Figure 9.13 Phase spectrum of a symmetrical FIR filter

The sinc function is not the only symmetrical impulse response. Figure 9.14 represents an alternative profile.

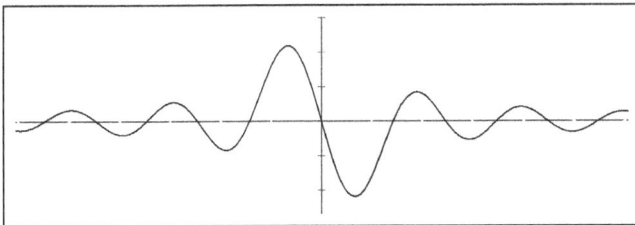

Figure 9.14: An alternative linear phase FIR filter

The profile in Figure 9.14 is said to have *negative symmetry* and occurs when there is a odd number of coefficients and the transfer function is,

$$H(\omega) = e^{-j\omega\Delta\left[\frac{N-1}{2}-\frac{\pi}{2}\right]} \sum_{n=1}^{n=(N-1)/2} a_n \sin(\omega n\Delta) \qquad ...9.16$$

There are other *symmetrical* FIR filters which display the characteristic of having a linear phase spectrum. The reason why this type of filter is important in audio applications lies in the phase integrity. Stereo and surround sound depend on the phase integrity of each audio channel. Each frequency is phase shifted by a fixed known amount. Whereas with other filter designs the phase can get seriously distorted with huge shifts at particular frequencies. This will be discussed in more detain in the next chapter on IIR Filters.

9.4 Commercial FIR Software

Probably the first commercial software for designing digital filters was *DFDP* (digital filter design pack) which was supplied by Texas Instruments. It was a DOS program and ran on the early IBM PC in the 1980s. On PCs containing an 8-bit Intel 8088 microprocessor it took several minutes to calculate the coefficients for a FIR filter using the Parks and McClellan Algorithm. The time was reduced slightly when Intel brought out the 8087 maths coprocessor. One interesting feature of this software was its code generation. At that time the popular DSP was the TMS320C10 and DFDP would quantise the coefficients and generate the assembly language code for the required filter. Since then digital filter design software has become easier to use, normally running under Microsoft Windows.

DADiSP

There is a FIR filter design facility which is an added extra to the full version of DADiSP and is part of the student version. As discussed in Chapter 1, the Filters icon in red is found on the top of the screen. We shall work through a couple of examples.

First generate a waveform which has two sine waves as shown in Figure 9.15.

W1: gsin(300,.001,56)-gsin(300,.001,211)

Also create the spectrum of the waveform in W1 as shown in Figure 9.16.

122

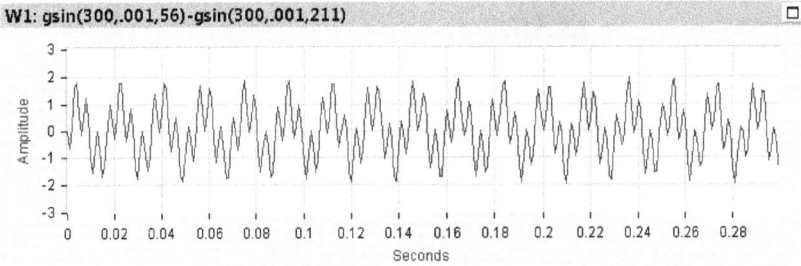

W1: gsin(300,.001,56)-gsin(300,.001,211)

Figure 9.15: A waveform with two sine waves

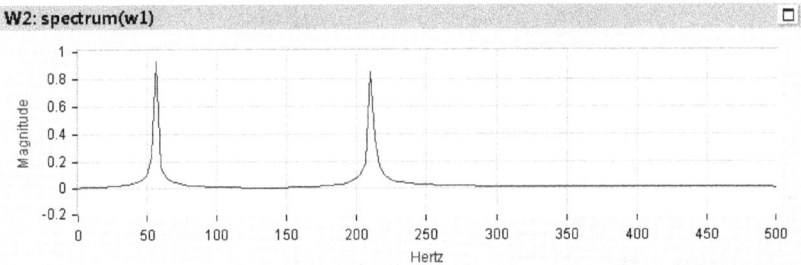

W2: spectrum(w1)

Figure 9.16: The spectrum of the waveform shown in W1

You will observe in Figure 9.16 the two spectral peaks, one at 56 Hz and the other at 211 Hz. We shall now design a FIR filter to remove the 211 Hz component from the waveform shown in Figure 9.15, this will in fact be a low pass filter. Click on **Filters** and select **FIR Filter Design** and the dialogue box as shown in Figure 9.17 will appear. You will observe various fields where appropriate specifications are entered. In this case, select,

- **Filter Type** → Low Pass
- **Filter Design** → Kaiser Window
- **Sampling Rate** → 1,000 Hz,
- **Cutoff Freq** → 120Hz
- **Stopband Edge** → 150Hz
- **Stopband Att** → 60dB
- **Destination Window** →W3.

123

The resulting response function is shown in Figure 9.18. To see the effect of this filter on the waveform in W1, select **Filters** again and choose **Filter Data** and specify the input data from W1 and output to W4. The results are shown in Figure 9.19 together with its spectrum in Figure 9.20.

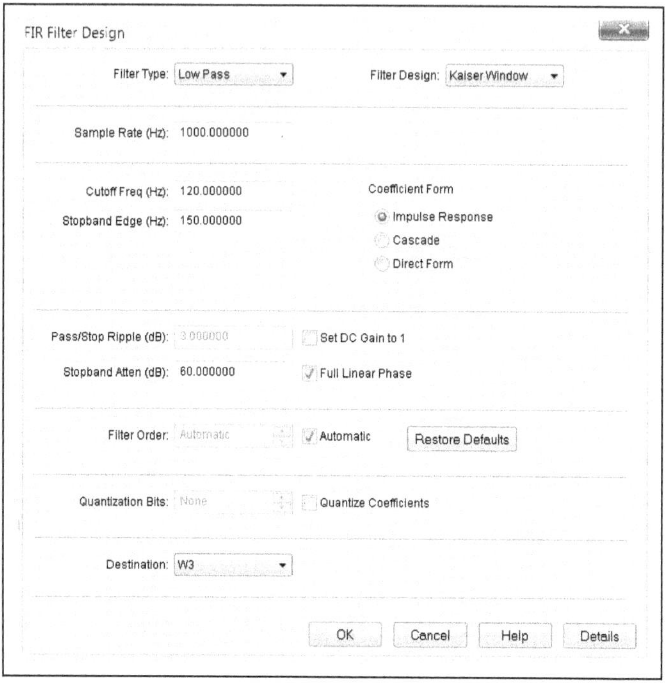

Figure 9.17: The FIR filter design dialogue box

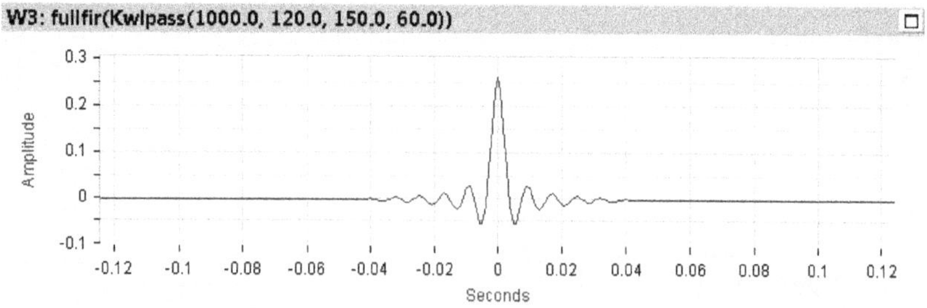

Figure 9.18: The sinc function generated from the filter design

W4: firfilterF(w1, W3)

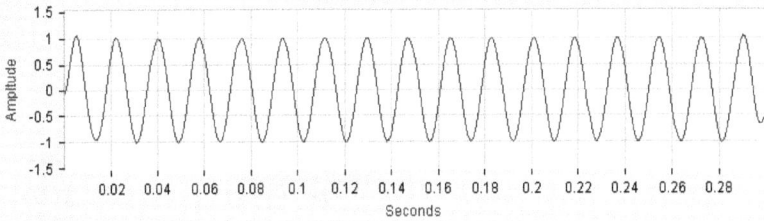

Figure 9.19: The filtered waveform from Figure 9.14

W5: spectrum(w4)

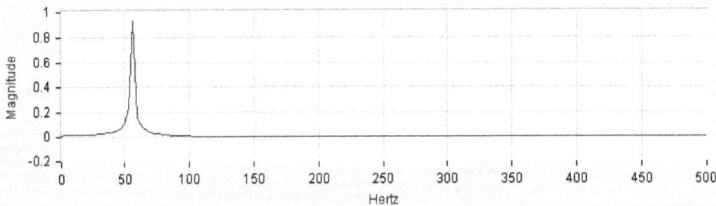

Figure 9.20: The spectrum of Figure 9.14 after filtering

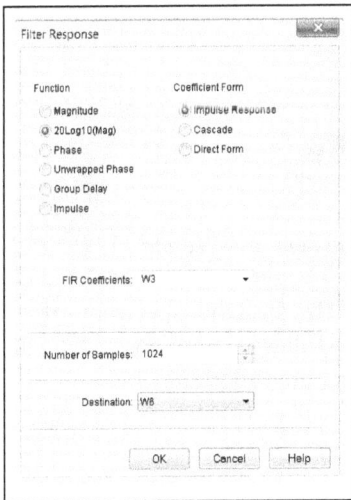

Figure 9.21: Features of the Filter Response

You will observe in Figure 9.20 the 211 Hz spectral component has been filtered out which clearly indicates the effectiveness of the filter. It is also instructive to see the transfer function for the filter. To achieve this, click on Filters and select Filter Response. The options are shown in Figure 9.21. Select 20log10(Mag) and Destination W6, the result is shown in Figure 9.22. In Figure 9.22 you will observe the presence of the many zeros in the stop band region of the filter. The effect of having so many zeros close together stops the transfer function rising above -119dB. In practise you would rarely have so many coefficients in a FIR filter. However it does demonstrate the overall effectiveness of the FIR low pass filter.

W6: 20*log10(filtmag(W3, [1], 1024))

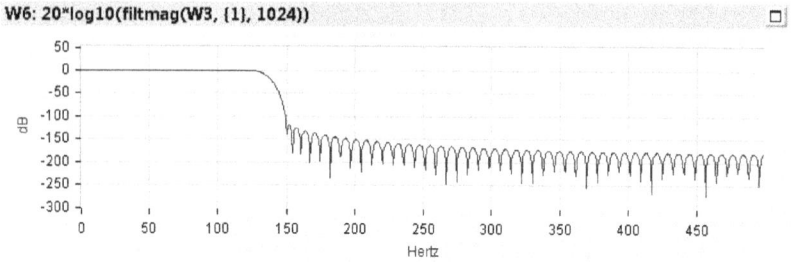

Figure 9.22: The transfer function of the filter response

High Pass FIR Filter Design

In this example we shall design a high pass filter and simulate the filter in operation. First, generate a waveform with two components.

W1: gsin(400,.001,47)-gsin(400,.001,172)

W2: spectrum(w1)

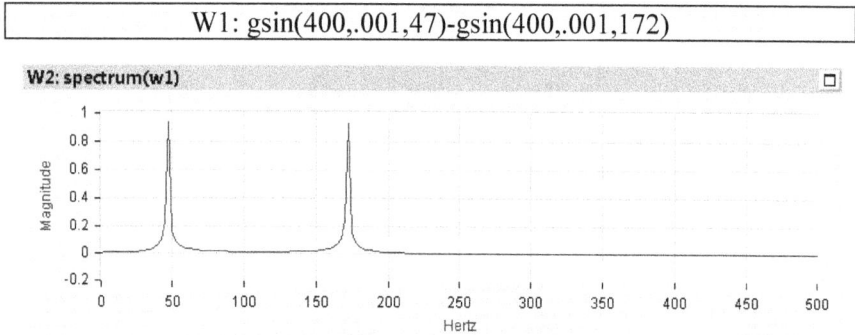

Figure 9.23: Spectrum of frequency component at 47Hz and 172 Hz

This has frequencies at 47 Hz and 172 Hz and the spectrum of this waveform is shown in Figure 9.23. Click on Filters and choose FIR Filter Design. When the dialogue box opens, select the following

- Filter Type → High Pass
- Filter Design → Remetz Exchange
- Sample Rate (Hz) → 1000.0
- Stopband Freq (Hz) → 120

- Cuttoff Freq (Hz) → 150
- Pass/Stop Ripple → 0.2dB
- Stopband Atten → 30 dB
- Destination → W2

The filter function coefficients appear in W2 as shown in Figure 9.24.

W2: Highpass(1000.0, 150.0, 0.2, 30.0, 120.0)

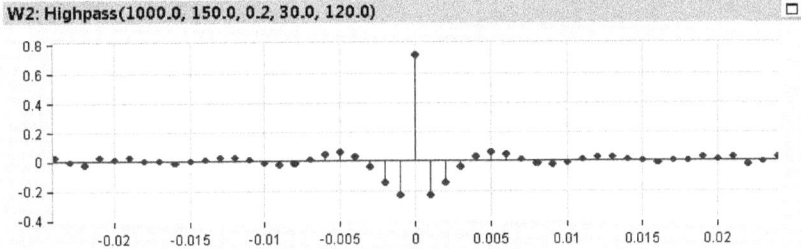

Figure 9.24: The coefficients for the high pass filter.

You will observe the coefficient's profile in Figure 9.24 depart from the normal sinc function. To obtain the transfer function for this filter, Click on Filters → Filter Response, select the following,

- 20Log10(Mag)
- FIR Coefficients → W2
- Number of Samples → 1024
- Destination → W3

The transfer function of the filter is shown in Figure 9.25 which has the expected characteristics of a high pass filter.

W3: 20*log10(filtmag(W2, {1}, 1024))

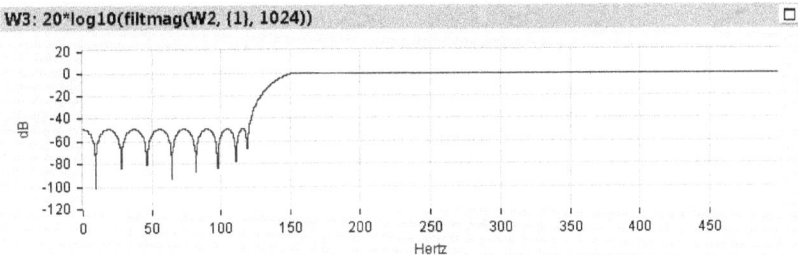

Figure 9.25: The transfer function for the high pass filter

127

To see the filter in action, select Filters → **Filter Data** and select the following,

- Input Data Series → W1
- Impulse Response
- Frequency Domain
- FIR Coefficients → W2
- Destination → W4

The spectrum of the resulting waveform is shown in Figure 9.26.

W5: spectrum(w4)

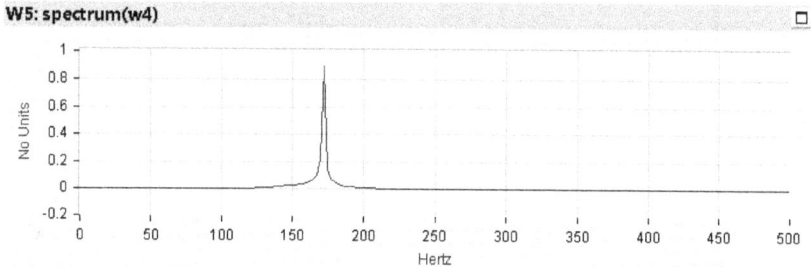

Figure 9.26: Spectrum of the filtered waveform in W4

You will observe from Figure 9.26 the low frequency component at 47 Hz has been removed which again demonstrates the effectiveness of a FIR filter. You will also note from the transfer function in Figure 9.25 the filter has far fewer zeros than in the previous example. The magnitude of the highest ripple value is -61.5dB (use the **Cursor** → **Crosshairs** to confirm this result). You are now encouraged to experiment with the FIR filter design features in DADiSP; it is good practice to print out your results. Use the **Preview Worksheet** icon to see your work before you print.

Other Commercial Software for FIR Design
There are a number of packages on the market and it's worthwhile to visit their websites to gain additional information. These include,

- ScopeFIR,   www.iowegian.com/scopefir
- WinFilter (free) www.winfilter.20m.com
- QEDesign 1000 www.kanecomputing.co.uk/mds_qedesign_1000.htm
- LabVIEW http://zone.ni.com/wv/app/doc/p/id/wv-216

Each have their merits and you should now be in a position to make judgements for yourself.

What you have gained from this chapter

1. The Window design of FIR filters.
2. The Hanning and Blackman Windows.
3. The Kaiser-Bessel Window with its shape factor.
4. An understanding of how windowing is a modulation.
5. Design specifications for a FIR filter.
6. The Equiripple FIR filter design.
7. The significance of symmetrical FIR filters
8. Commercial software for FIR filter design.

The DADiSP skills you have acquired from this Chapter

1. *gsinc* - to generate a sinc function
2. *gimpulse* - to generate an impulse function
3. *fconv* - to obtain a convolution function
4. *20lot10(spectrum)* - to obtain a dB scale spectrum
5. *ghanning* - to generate a Hanning Window
6. *overlay* - to overlay one window with another
7. *gblackman* - to create a Blackman Window
8. *gkaiser* - to generate a Kaiser-Bessel Window
9. *fulfil, firfiltf & filtag* - part of the DADiSP filter design

10. Infinite Impulse Response (IIR) Filters

The main difference between the FIR filter and an Infinite Impulse Response (IIR) filter is the latter has *recursion*. By this we mean its current output is dependent on the previous outputs. The output is fed back into the filter. The general expression for an IIR filter is,

$$y(m) = \sum_{n=0}^{n=N-1} b_n x(m-n) - \sum_{k=1}^{k=K-1} a_k y(m-k) \qquad \text{... 10.1}$$

In addition to the FIR filter terms there are the recursive components relating to the previous outputs. You will observe a new set of coefficients $\{a\}$. To illustrate the recursive concept consider the case when $N = 3$ and $K = 2$,

$$y(m) = b_0 x(m) + b_1 x(m-1) + b_2 x(m-1)$$
$$- a_1 y(m-1) + a_2 y(m-2) \qquad \text{... 10.2}$$

From this expression a schematic can be constructed as seen in Figure 10.1.

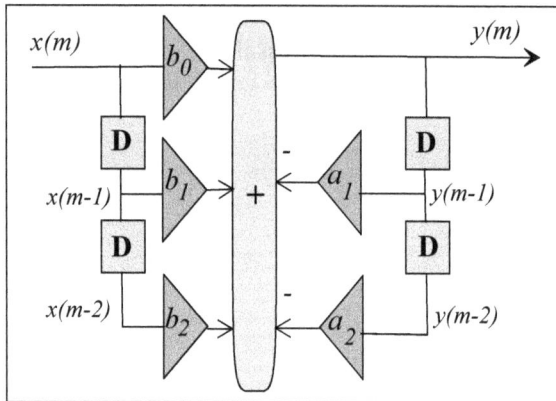

Figure 10.1: A schematic showing the concept of feedback

As can be seen from Figure 10.1, the delayed outputs $y(m-1)$ and $y(m-2)$ are fed back into the filter. You will also notice there are four memory elements (**D**). An obvious question is why is it called an *infinite* impulse response filter. To answer this question consider the following filter,

$$y(m) = bx(m) - ay(m-1) \qquad \text{... 10.3}$$

We shall follow the filter through for the first few samples; when this filter starts ($m = 0$), then

$$y(0) = bx(0)$$

$$y(1) = bx(1) - ay(0) = bx(1) - a[bx(0)]$$

$$y(2) = bx(2) - ay(1) = bx(2) - a[bx(1) - a[bx(0)]]$$

$$y(3) = bx(3) - ay(2) = bx(3) - a[bx(2) - a[bx(1) - a[bx(0)]]]$$

... 10.4

You can appreciate that *y(3)* has a contribution from *x(3)*, *x(2)*, *x(1)* and *x(0)*. By the time have the Mth sample, *y(M)* has a contribution from every previous *M-1* input samples. So if *M* was 1,000,000, *y(M)* would have a contribution from the current millionth sample and the previous 999,999 samples - hence it's called an infinite because it retains a contribution from every single input sample. Somewhat similar to *sherry*. For every new cask of sherry it is customary to pour a bottle of last year's sherry into it. So from a given vineyard, every bottle contains some sherry that has ever been produced from that vineyard. Incidentally IIR filters are closely related to active analogue filters which if you recollect are made with capacitors which act as the memory for the filter. In fact IIR filters are designed from analogue filters which is discussed later.

10.1 The Concept of the Pole

Have glanced at IIR filters in the time domain, their performance in the Z domain is of huge importance. We therefore perform a Z-transform on Eq:10.1.

$$Z\{y(m)\} = Z\{\sum_{n=0}^{M-1} b_n x(m-n)\} - Z\{\sum_{k=1}^{K-1} a_k y(m-k)\} \qquad \text{... 10.5}$$

This becomes,

$$Y(z) = X(z) \sum_{n=0}^{N-1} b_n z^{-n} - Y(z) \sum_{k=1}^{K-1} a_k z^{-k} \qquad \text{... 10.6}$$

Rearranging,

$$Y(z)\left[1 + \sum_{k=1}^{K-1} a_k z^{-k}\right] = X(z) \sum_{n=0}^{N-1} b_n z^{-n} \qquad \text{... 10.7}$$

The transfer function is

$$H(z) = \frac{Y(z)}{X(z)} = \frac{\displaystyle\sum_{n=0}^{N-1} b_n z^{-n}}{1 + \displaystyle\sum_{k=1}^{K-1} a_k z^{-k}} \qquad \text{... 10.8}$$

When the sums are factorised we get,

$$H(z) = \frac{\displaystyle\prod_{n=1}^{N-1}(z - z_n)}{\displaystyle\prod_{k=1}^{K-1}(z - p_k)} \qquad \text{... 10.9}$$

Eq:10.9 leads to the concept of a pole. The numerator refers to the locations of the zeros with which we are already familiar. It is the behaviour of the denominator which is of great interest. The value p_k is usually complex and they, like zeros occur in conjugate pairs. What happens to the transfer function $H(z)$ whenever $z = p_k$? Well you can see that when this happens $H(z) \rightarrow \infty$. This is the nature of a *pole* and these are akin to a resonance. The effect of the pole depends on how close it resides to the perimeter of the unit circle and its location is designated by the $\times$ symbol. To gain a further understanding of a pole we shall consider a damped sine wave. In DADiSP construct a damped sine wave as shown in Figure 10.2.

W1: gsin(300,.01,12)*gexp(300,.01,-2)

The equation for a damped complex sine wave is,

$$x(t) = A_0 e^{-\alpha t} e^{j\omega_o t} \qquad \text{... 10.10}$$

which may be expressed as,

$$x(t) = A_o e^{j(\omega_o + j\alpha)t} \qquad \text{... 10.11}$$

where A_o is its maximum amplitude. The degree of damping on the sine wave in Figure 10.2 depends on the value of α and this is significant when considering the location of the poles in the unit circle.

```
W1: gsin(300,.01,12)*gexp(300,.01,-2)                    □
```

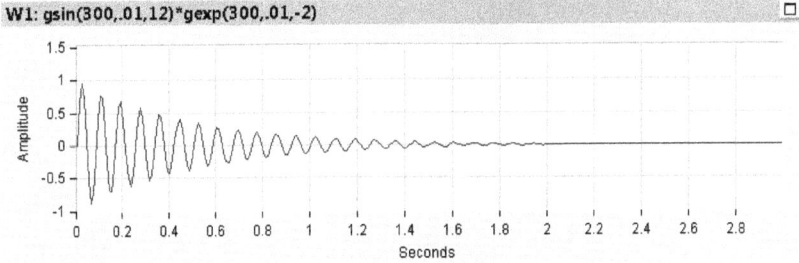

Figure 10.2: A damped sine wave

Now perform a Fourier Transform on Eq: 10.11,

$$F(\omega) = A_o \int_0^\infty e^{j(\omega_o+ja)t}e^{-j\omega t}dt = A_o \int_0^\infty e^{-j(\omega-\omega_0+ja)}dt$$

$$= A_o \left[\frac{e^{-j(\omega-\omega_0+ja)t}}{-j(\omega-\omega_0+ja)} \right]_0^\infty$$

... 10.12

Expand the square brackets,

$$F(\omega) = -j\frac{A_o}{(\omega-\omega_o+ja)}$$

... 10.13

The square of the magnitude of the spectrum is therefore,

$$|F(\omega)|^2 = \frac{A_o^2}{(\omega-\omega_o)^2+a^2}$$

... 10.14

To see what this function looks like, perform a *spectrum* on the waveform in W1, the spectrum is shown in Figure 10.3.

```
W2: spectrum(w1)
```

Eq:10.14 represents a broadened spectral line; the greater the value of α, the greater the degree of spectral broadening. Figure 10.3 shows a damped resonance which can be characterised by its Q *value* which is defined as,

$$Q = \frac{f_0}{f_2-f_1}$$

... 10.15

where f_o is the central frequency and $f_2 - f_1$ is the spectral width at *half height*.

W2: spectrum(w1)

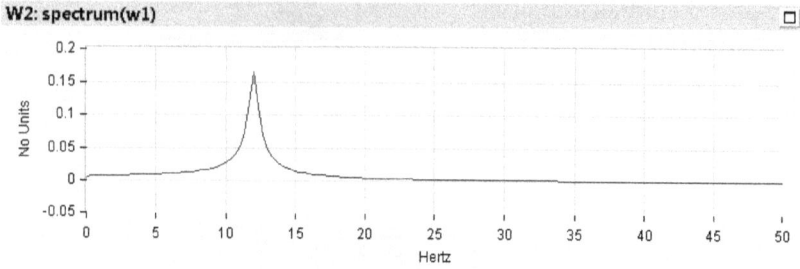

Figure 10.3: Spectrum of a damped sine wave

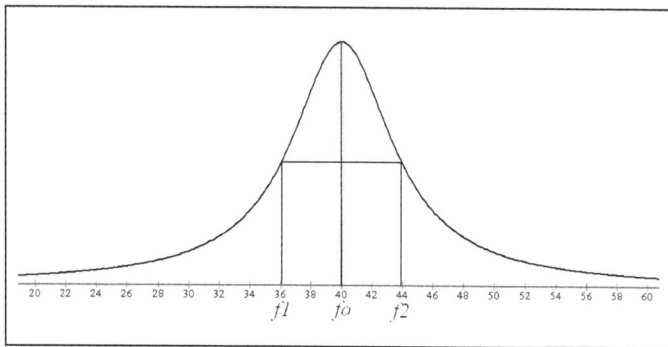

Figure 10.4: A broadened resonance.

Consider the resonance in Figure 10.4 the greater the damping, the larger the value of $f_2 - f_1$ and the smaller the value of Q. The Q value of the resonance in Figure 10.4 is 40/(44 - 36) = 5. In effect Q measures the sharpness of the resonance and for any pole in a IIR filter, its Q will depend on how *close it is the perimeter* of the unit circle.

10.2 Poles and the Unit Circle

Having discussed poles and how they arise, it's time to see them in the context of the unit circle. Consider the following IIR filter difference equation,

$$y(m) = x(m) + 0.5y(m-1) - 0.95y(m-2) \qquad \dots 10.16$$

It behaviour in the Z domain is achieved by performing a Z-transform on the expression,

$$Z\{y(m)\} = Z\{x(m)\} + 0.5Z\{y(m-1)\}$$
$$-0.95Z\{y(m-2)\} \qquad \text{... 10.17}$$

Leaving,

$$Y(z) = X(z) + 0.5Y(z)z^{-1} - 0.95Y(z)z^{-2} \qquad \text{... 10.18}$$

Rearranging to give the transfer function,

$$H(z) = \frac{1}{1 - 0.5z^{-1} + 0.95z^{-2}} = \frac{z^{-2}}{z^2 - 0.5z + 0.95} \qquad \text{... 10.19}$$

$$= \frac{z^{-2}}{(z - 0.25 + j0.942)(z - 0.25 - j0.942)}$$

This expression has poles at $z = 0.25 + 0.942j$ and $z = 0.25 - 0.942j$. We now use DADiSP to investigate this transfer function. First create an impulse.

W1: gimpulse(300,.01,0)

The filter response function can be generated from Eq:10.16 by

W2: Filteq({1}, {.5,-.95}, W1)

You will notice the actual coefficients from Eq:10.16 are used. The result of using this command is shown in Figure 10.5

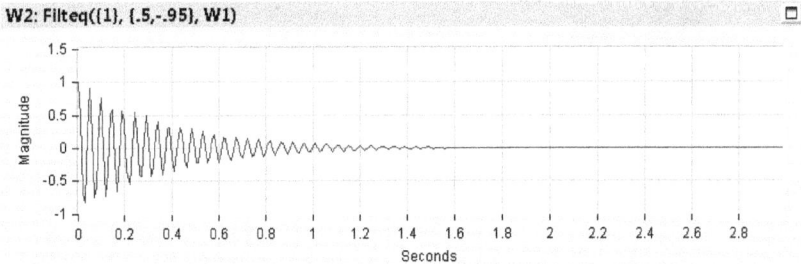

Figure 10.5: The impulse response for Eq:10.16.

DADiSP also has a command for showing the unit circle and the result is shown in Figure 10.6.

W4: zplane({1}, {1,-.5,.95})

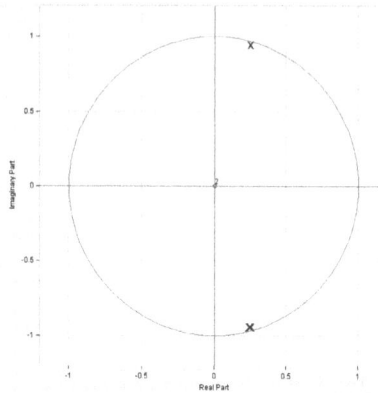

Figure 10.6: The unit circle showing the poles of Eq:10.15

The poles being so close to the perimeter of the unit circle suggests there must be a significant resonance. This can be deduced from Figure 10.5 showing a damped response. The transfer function of this response is shown in Figure 10.7.

W3: spectrum(w2)

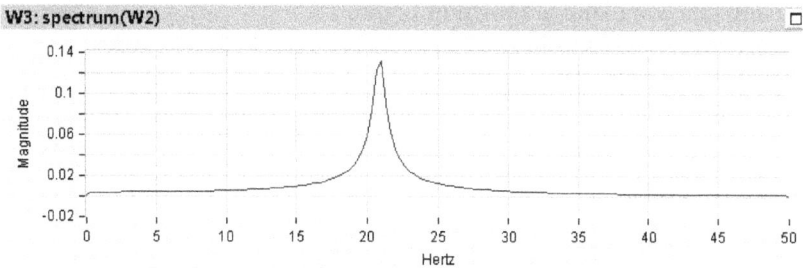

Figure 10.7: Spectrum of the IIR filter response of Eq:10.15

The resonance frequency can be calculated from a root in Eq:10.19. The root in the upper half of the unit circle is $z = 0.5+0.942j$. Convert this into polar coordinates, $|z| = 0.97$ and $\theta = 75.136°$. Half the sampling frequency *fs/2* is *50Hz*, therefore *50Hz* ≡ *180°*, the frequency of the resonance is given by,

$$(\tfrac{50}{180}) \times 75.136 = 20.87\text{Hz}$$

which agrees with the spectrum in Figure 10.7. There is a command in DADiSP which allows you to examine the performance of a digital filter directly from its Z domain expression. For example, consider the filter,

$$H(z) = \frac{1-0.4z^{-1}+0.9z^{-2}}{1-0.6z^{-1}-0.4z^{-2}+0.8z^{-3}} \qquad \dots 10.20$$

The command is,

W3: freqz({1, - 0.4, 0.9}, {1, -0.6, -0.4, 0.8}, 1024, 500)

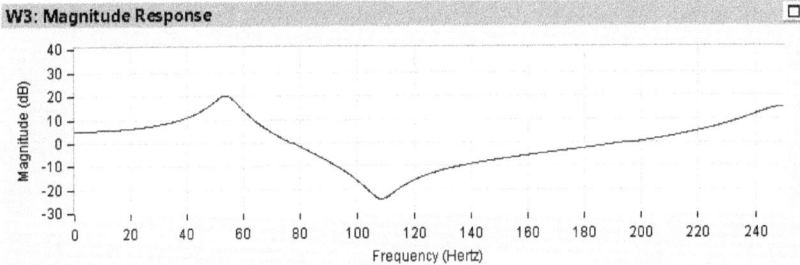

Figure 10.8: Magnitude response of Eq:10.20

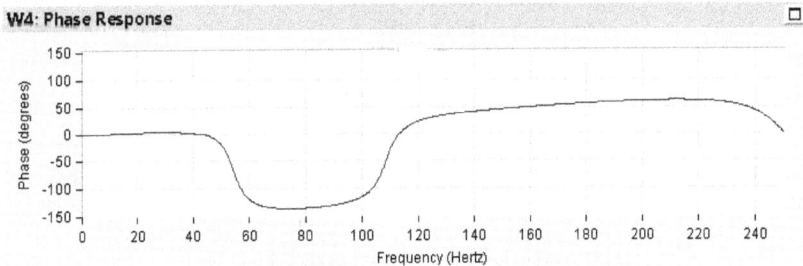

Figure 10.9: Phase response of Eq:10.20

As can be seen from Figures 10.8 and 10.9 the magnitude and phase spectral are produced in consecutive windows. This is a very convenient command for gaining an insight into the behaviour of any filter. You will observe from Figure 10.8 the magnitude response has a pole and a zero.

The positions of the poles and zeros in the Z domain can be determined by using the DADiSP Zplane command and the result is shown in Figure 10.10.

W5: freqz({1, - 0.4, 0.9}, {1, -0.6, -0.4, 0.8})

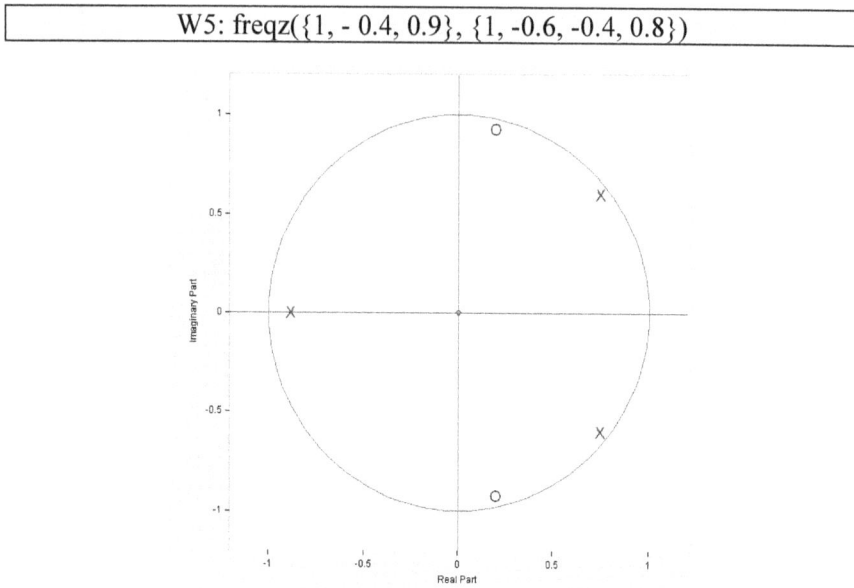

Figure 10.10: The unit circle showing the locations of the poles and zeros in Eq: 10.20

10.3 Biquadratic Realisation

When looking at Figure 10.1 you will observe the filter requires four memory elements. It is possible to reduce the number of memory elements and this is referred to as the *biquadratic realisation*. Start with a second order IIR filter,

$$y(n) = b_o x(n) + b_1 x(n-1) + b_2 x(n-1)$$
$$- a_1 y(n-1) - a_2 y(n-2)$$

... 10.21

Perform a Z-transform to obtain,

$$\frac{Y(z)}{X(z)} = \frac{(b_o + b_1 z^{-1} + b_2 z^{-2})}{(1 + a_1 z^{-1} + a_2 z^{-2})}$$

... 10.22

Now introduce a new variable *U(z)* so that,

$$\frac{Y(z)}{X(z)}\frac{U(z)}{U(z)} = \frac{Y(z)}{U(z)}\frac{U(z)}{X(z)} = \frac{(b_0+b_1z^{-1}+b_2z^{-2})}{(1+a_1z^{-1}+a_2z^{-2})} \qquad \text{... } 10.23$$

Now let,

$$\frac{Y(z)}{U(z)} = (b_0 + b_1z^{-1} + b_2z^{-2}) \text{ and } \frac{U(z)}{X(z)} = \frac{1}{1+a_1z^{-1}+a_2z^{-2}} \qquad \text{... } 10.24$$

These can be expressed as,

$$Y(z) = U(z)b_0 + b_1U(z)z^{-1} + b_2U(z)z^{-2}$$
$$X(z) = U(z) + a_1U(z)z^{-1} + a_2U(z)z^{-2} \qquad \text{... } 10.25$$

Now perform the inverse Z-transform on these expressions,

$$y(n) = b_o u(n) + b_1 u(n-1) + b_2 u(n-2)$$
$$u(n) = x(n) - a_1 u(n-1) - a_2 u(n-2) \qquad \text{... } 10.26$$

Schematics can be constructed for these expressions as shown in Figure 10.11. In this you will see the two sections corresponding to the two equations in Eq:10.26 and the positioning of the new variable *u(n)*. The memory elements for *u(n-1)* and *u(n-2)* are common to both sections.

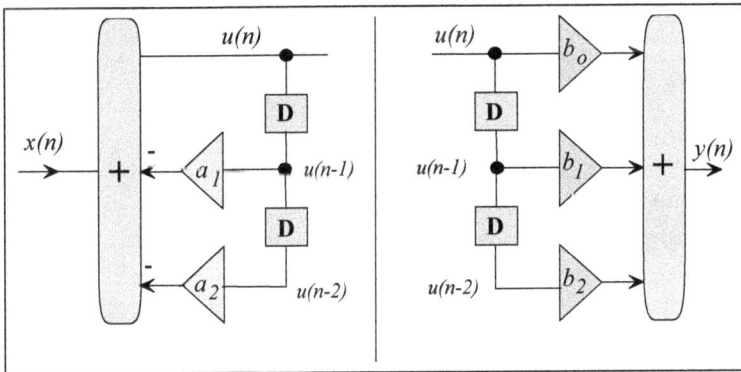

Figure 10.11: Schematic representations of Eq:10.26

There is no reason why the two sections cannot be combined as shown in Figure 10.12. The schematic shown in Figure 10.12 is known as the *canonical biquadratic realisation* or *biquadratic section* (BQS) which is distinguished by only having two memory elements as opposed to four. Implementing IIR filters as biquadratic sections reduces the memory demand by ½. When coding the filter, Eq:10.21 would not be used, instead

Eq:10.26 would be used introducing the intermediate variable *u(n)*. Each BQS is a second order structure with two memory elements, five multiplications and five additions an you can see in Figure 10.12 how these different elements fit together to form a BQS. This structure is fundamental in the realisation and operation of IIR filters.

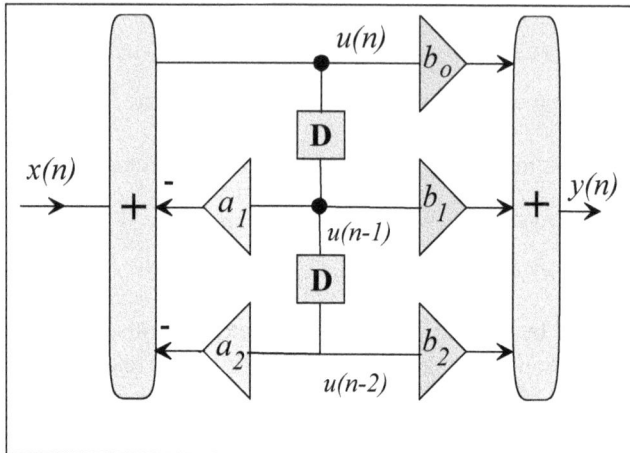

Figure 10.12: Combining the two sections in Figure 10.21

10.4 Calculating the Effect of the Poles and Zeros

Although the unit circle is a useful device for providing a visual representation of the locations of the poles and zeros, their actual effect can also be calculated from the unit circle. Consider the position of two poles and two zeros in Figure 10.13. Take an arbitrary position on the perimeter of the unit circle S, a vector is drawn to the two poles and two zeros. The influence of these on the location at S is,

$$H(z)_S = \frac{z_1 z_4}{p_2 p_3} = \frac{|z_1||z_4|}{|p_2||p_3|} e^{j(\theta_1+\theta_4-\theta_2-\theta_3)} \qquad \text{... 10.27}$$

This may be expressed as,

$$|H(z)|_S = \frac{|z_1||z_4|}{|p_2||p_3|} \text{ and } \phi(z)_S = \theta_1 + \theta_4 - \theta_2 - \theta_3 \qquad \text{... 10.28}$$

You can appreciate from this expression that as the *locus* S approaches a pole, $|H(z)|$ gets larger since $|p2|$ or $|p3|$ get smaller. Likewise as locus S gets closer to the zeros $|H(z)|$ diminishes in size. You can also appreciate how the phase $\phi(z)$ also varies as the locus S moves round the perimeter of the unit circle.

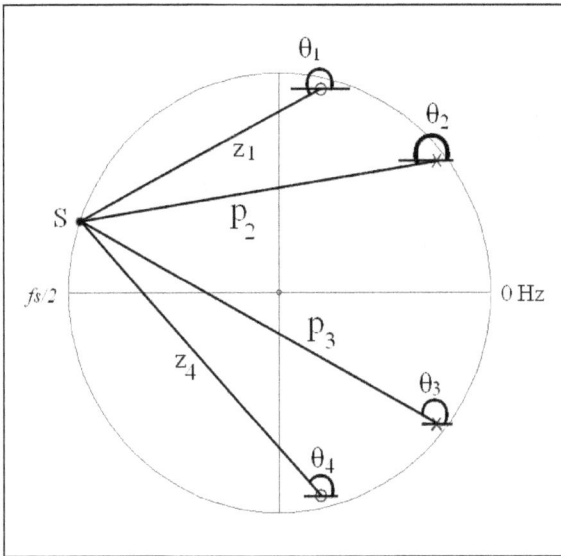

Figure 10.13: Locations of poles and zero in the unit circle

10.5 Realising High Order IIR Filters

When presented with an Nth order filter, by using biquadratic realisations, it would be represented as $N/2$ biquadratic sections normally in series. Figure 10.14 shows an example of a 10th order filter. This has five Biquadratic sections (BQS).

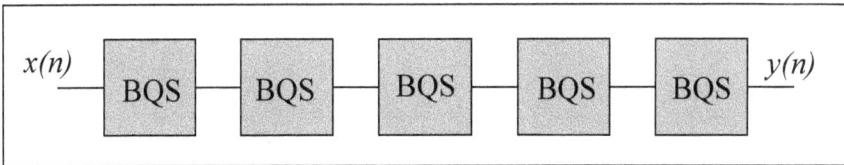

Figure 10.14: A 10th order filter realised as five BQSs

One way of thinking of this structure operating is to consider the signal entering on the left and as it propagates through each section it becomes more and more refined until it emerges on the right with all the unwanted frequencies removed. The design shown in Figure 10.14 is a serial configuration, although a parallel configuration can also be realised. In Chapter 11 you will find out how commercial software displays the coefficients for IIR filters - usually in sets of 5 coefficients for each BQS.

10.6 Digital Resonator

It has been stated that when a pole passes outside the unit circle for whatever reason, the IIR filter becomes unstable and the response function grows exponentially. However when a poles resides on the perimeter of the unit circle the response function will oscillate. Its resonance frequency depends where on the perimeter the pole resides. Consider the following,

$$H(z) = \frac{z^2}{(z-p_1)(z-p_1^*)} \qquad \dots 10.29$$

If the conjugate poles reside on the perimeter of the unit circle then, $|p_1| = 1$, in which case,

$$H(z) = \frac{z^2}{(z-e^{j\theta})(z-e^{-j\theta})} = \frac{z^2}{z^2 - z(e^{j\theta}+e^{-j\theta})+1} \qquad \dots 10.30$$

Eq:10.30 can be expressed as,

$$H(z) = \frac{1}{1-2z^{-1}\frac{(e^{j\theta}+e^{-j\theta})}{2}+z^{-2}} = \frac{1}{1-\cos(\theta)z^{-1}+z^{-2}} \qquad \dots 10.31$$

Performing the inverse Z-transform on 10.31,

$$y(n) = x(n) + 2\cos(\theta)y(n-1) - y(n-2) \qquad \dots 10.32$$

which is in effect a digital resonator where θ (measured in radians), determines its frequency. To illustrate this resonator, select a sampling frequency of 1,000 Hz. For it to oscillate at 150 Hz,

$$500\text{Hz} \equiv \pi, \text{ and } 150\text{Hz} \equiv x, \text{ or } x \equiv 15\pi/50 = 0.942 \text{ radians.}$$

Therefore $2\cos(0.942) = 1.1763$. The difference equation becomes,

$$y(n) = x(n) + 1.1763y(n-1) - y(n-2)$$

In DADiSP create an impulse,

W1: gimpulse(200, .001)

Create the filter and observe the output as shown in Figure 10.24,

---

W2: filteq({1}, {1.1763, -1}, w1)

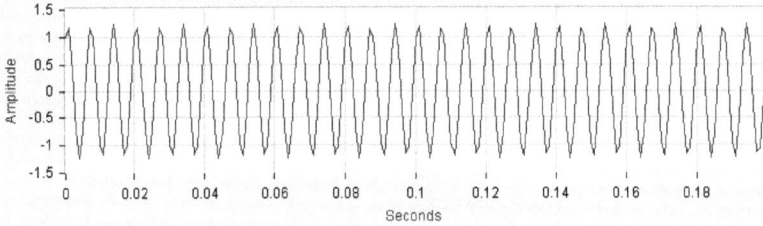

Figure 10.15: A resonating IIR filter

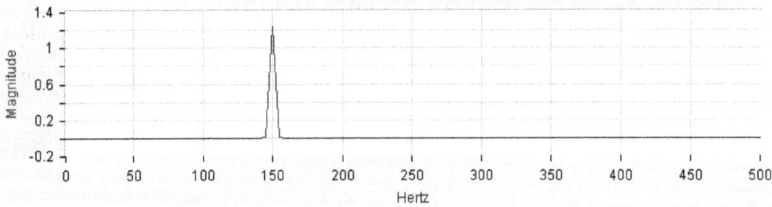

Figure 10.16: The spectrum of the resonance

The spectrum of the resonance is shown in Figure 10.16 and it should come as no surprise the resonance occurs at 150 Hz as calculated. All that is needed to start the resonance is a single impulse after that the resonance is self sustaining. A word of caution, when attempting to implement this resonator on a fixed point (FX) digital signal processor (DSP), the resonator is recursive and the round off effects are likely to cause the poles to drift outside the unit circle. If this should happen it will lead to exponential growth in the amplitude and the failure of the IIR resonator to be useful.

What you have gained from this Chapter

1. The difference equation representing an IIR filter.
2. The schematic of an IIR filter.
3. Why IIR filters are referred to as infinite.
4. The concept of a pole.
5. The spectral profiles of a damped sine wave.

6. Characterising a spectral line with a Q-factor.
7. Poles and the unit circle.
8. Biquadratic structures.
9. Calculating the effects of poles and zeros.
10. Realising higher order IIR filters.
11. Constructing a digital sine wave resonator.

The DADiSP skills you have acquired from this Chapter

1. *gsine* - to create a sine wave
2. *gexp* - to create an exponential function
3. *spectrum* - to obtain a spectrum
4. *gimpulse* - to create an impulse function
5. *Filteq* - to create the impulse response of a filter
6. *Zplane* - to view the positions of poles and zeros in the unit circle
7. *Freqz* - to obtain the magnitude and phase spectra in the frequency domain from the Z-domain.

DADiSP Extra

There are several special waveforms you can generate in DADiSP, click on the Function Wizard $f_x \to$ Generate Data $\to$ Other and in the *Generate Miscellaneous Functions* dialogue box which opens, click on Series Type and select RTSQR with a frequency of 4 and the waveform in Figure 10.17 will appear.

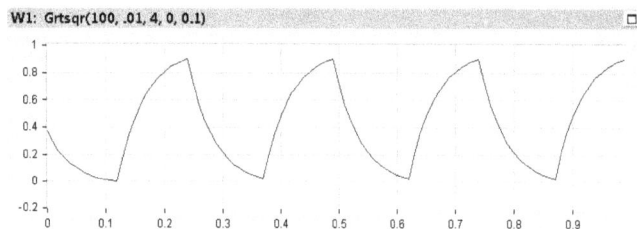

Figure 10.17: The RTSQR function

This is common waveform found in electronic circuits especially where the charging up and discharging of capacitors are involved - often seen in the operation of *comparators*.

11. Designing Infinite Impulse Response Filters

Many IIR filters are actually modelled on analogue filters. Four of the most well known designs are Butterworth, Chebyshev (I & II) and Elliptical and the transfer functions for these filters are shown in Figure 11.1. Each has slightly different characteristics and their selection depends on the specifications of the filter. You will observe that three of them have ripple and out of the four, the greatest roll-off can be achieved with an Elliptical design which has a minimum number of coefficients. However the cost is the severe effect the filter has on the phase characterises of the signal. This filter also has ripple in both the band pass and band stop regions. In a similar manner to the FIR filter the coefficients are derived from digitising the impulse response of the filter.

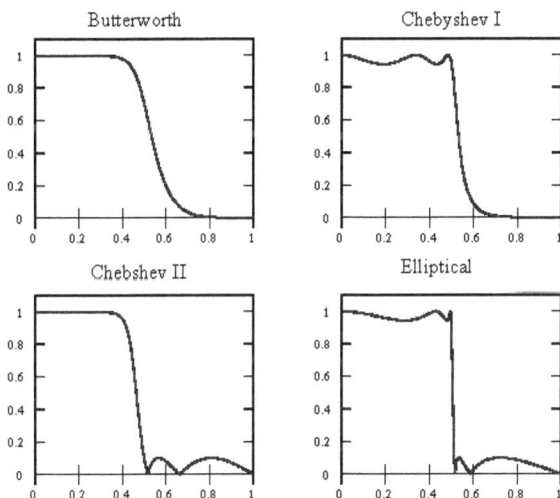

Figure 11.1: The transfer functions of four well known analogue filters

In IIR filters the main emphasis is on the positioning of the poles and not so much on the zeros. To illustrate this point consider the following IIR filter,

$$H(z) = \frac{(z-0.2)}{(z+0.5)(z-0.3)} \qquad \ldots 11.1$$

Given a factorised expression, such as Eq:11.1, if the order of the denominator is greater than the order of the nominator, it can be expressed as *partial fractions*. Eq:11.1 becomes,

$$H(z) = \frac{0.875}{(z+0.5)} - \frac{0.125}{(z-0.2)} \qquad \ldots 11.2$$

Further information on partial fractions is given in the Appendix where there are examples of using DADiSP to perform the calculations. You will observe from Eq:11.2 there is no longer a zero. When digitising the impulse response of an IIR filter a number of difficulties arise when putting the filter into practise. The digital variant of the filter does not perform in the same manner to that of its analogue original. This is particularly true when approaching half the sampling frequency.

11.1 Impulse Invariant Design

To gain an insight into how analogue filters are modelled it is customary for engineers to use the Laplace S plane. Consider the transfer function of a single pole filter,

$$H(s) = \frac{C}{s-p} \qquad \ldots 11.3$$

The equivalent transfer function of the sampled version of this filter in the Z domain is given by,

$$H(z) = \frac{C}{1-e^{p\Delta}z^{-1}} \qquad \ldots 11.4$$

If the original filter has N poles, in the S domain it's represented as,

$$H(s) = \sum_{k=1}^{k=N} \frac{C_k}{s-p_k} \qquad \ldots 11.5$$

The digital equivalent of this expression is,

$$H(z) = \sum_{k=1}^{k=M} \frac{C_k}{1-e^{p_k\Delta}z^{-k}} \qquad \ldots 11.6$$

The impulse response of the discrete filter is identical to the original when sampled at intervals of Δ. It is therefore called an *impulse invariant* (not changing) filter. However in the frequency domain, the digital equivalent filter profile repeats itself in multiples of the sampling frequency *fs* as shown in Figure 11.2.

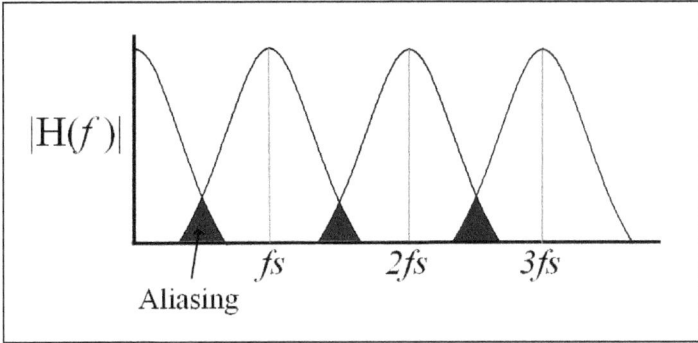

*Figure 11.2: The digitised transfer function of
the filter repeating itself with a frequency of fs Hz*

One way of looking at the digitised transfer in Figure 11.2 is rotating round the unit circle in an anticlockwise direction, each time you pass *fs/2* a new cycle is started. A problem with the impulse invariant design is the transfer function profile often extends beyond *fs/2* which causes an overlap with the next profile as seen in Figure 11.2. This gives rise to aliasing - the spurious appearance of unwanted frequencies in the sampled data. A method is therefore required to prevent this from happening.

11.2 Bilinear Transform

To constrain the transfer function profile within the limits $f \in [-fs/2, +fs/2]$ a *mapping* device is used. Given a transfer function for an analogue filter in the S domain, *s* is replaced by,

$$s \to \frac{z-1}{z+1} \qquad \text{... 11.7}$$

An interesting question, why should the mapping take on this format? To answer this, if *z* is replaced by $e^{j\omega\Delta}$, then

$$\frac{z-1}{z+1} \to \frac{e^{j\omega\Delta}-1}{e^{j\omega\Delta}+1} = \frac{e^{j\omega\Delta/2}-e^{-j\omega\Delta/2}}{e^{j\omega\Delta/2}+e^{-j\omega\Delta/2}} = j\tan(\frac{\omega\Delta}{2}) \qquad \text{... 11.8}$$

This tan mapping process is shown in Figure 11.3. You will note that at the two limits *-fs/2*, tan($\omega\Delta/2$) $\to -\infty$ and at *+fs/2*, tan($\omega\Delta/2$) $\to +\infty$. The transfer function is therefore constrained between these two limits. In Figure 11.3 the analogue transfer function is shown on the left and this is mapped to a constrained digital transfer function indicated by the dotted line (·····).

A point on the analogue transfer function A maps to a point A' on the digitised transfer function.

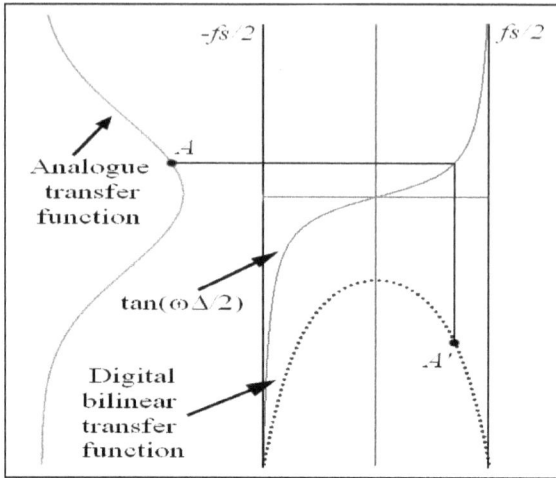

Figure 11.3: The tan($\omega\Delta/2$) mapping function

You will observe the digitised transfer function is constrained within $-fs/2$ and $+fs/2$. However the mapping has a problem - it warps the transfer function. It is therefore common practise to *pre-warp* the transfer function before the mapping. To demonstrate how the pre-wrap works, consider an example. An analogue low pass filter has the following transfer function,

$$H(s) = \frac{1}{s^2 + \sqrt{2}\,s + 1} \qquad \ldots 11.9$$

The sampling frequency is 10 kHz and the 3dB cut-off is at 2 kHz. The sampling interval $\Delta = 10^{-4}$s. The pre-wrap critical frequency is,

$$\omega_p' = \tan(\omega_p \tfrac{\Delta}{2}) = \tan\left(\frac{2\pi \times 2000 \times 10^{-4}}{2}\right) = \tan(\tfrac{\pi}{5}) = 0.7265 \qquad \ldots 11.10$$

Therefore the pre-wraped transfer function becomes,

$$H(s) = \frac{1}{\left(\frac{s}{\omega_p'}\right)^2 + \sqrt{2}\left(\frac{s}{\omega_p'}\right) + 1} = \frac{\omega_p'^2}{s^2 + \sqrt{2}\,\omega_p's + \omega_p'}$$

$$= \frac{0.5278}{s^2 + 1.027s + 0.5278} \qquad \ldots 11.11$$

Now make the substitution given in Eq:11.7

$$H(z) = \frac{0.5278}{\left(\frac{z-1}{z+1}\right)^2 + 1.027\left(\frac{z-1}{z+1}\right) + 0.5278} \qquad \dots 11.12$$

or

$$H(z) = \frac{0.5278z^2 + 1.0556z + 0.5278}{2.5548z^2 - 0.9440z + 0.5008} \qquad \dots 11.13$$

Performing the inverse Z-transform on this expression we arrive at the difference equation,

$$y(n) = 0.2066x(n) + 0.4131x(n-1) + .2066x(n-2)$$
$$+ 0.3696y(n-1) - 0.1960y(n-2)$$
$$\dots 11.14$$

Given any analogue transfer function, by performing the pre-warp frequency substitution and replacing the s according to Eq:11.7 you obtain the digital equivalent IIR filter. To see the performance of this IIR filter in DADiSP, first generate an impulse with sample separation of 10^{-4}s. This is equivalent to a sampling frequency of 10 kHz

W1: gimpulse(600,.0001)

Now create the filter with the coefficients taken from Eq:11.14,

W2:Filteq({.2066, .4131, .2065}, {.3696, -.1960}, w1)

The transfer function is obtained from the spectrum of w2 which is shown in Figure 11.4. This has been scaled by 300.03 to give the pass band a magnitude of 1.

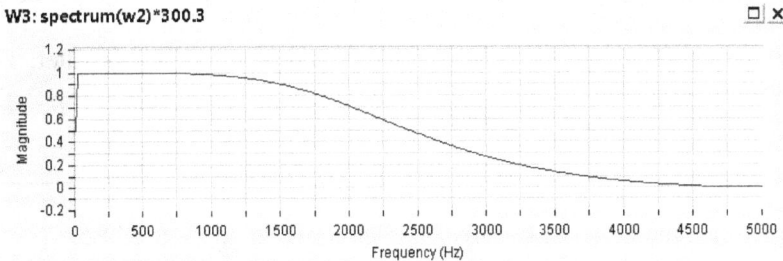

Figure 11.4: The transfer function of the filter in Eq:11.12

Clearly Figure 11.4 is a low pass filter and it can be estimated that the 3 dB point (reduction in 50%) is 2.424 kHz (you can zoom in on the 0.5 cross over point). As you can imagine for higher order filters the transfer function become far more complicated. However commercial software for designing IIR filters performs all the calculations. Pre-warping is also performed automatically.

When using commercial design software you are usually given a choice of filter option depending upon the specifications. We shall work through another example. The transfer function of the filter is,

$$H(s) = \tfrac{1}{s+1} \qquad \text{... 11.15}$$

The sampling frequency is 150 Hz and the cut-off frequency is 30 Hz, design an equivalent bilinear digital filter. First calculate the pre-warp frequency,

$$\omega_p' = \tan\!\left(\tfrac{\omega_{p\Delta}}{2}\right) = \tan\!\left(\tfrac{2\pi \cdot 30}{150/2}\right) = 0.7265 \qquad \text{... 11.16}$$

Now pre-warp the filter by replacing s by s/ω_p,

$$H(s') = \tfrac{1}{\frac{s}{0.7265}+1} = \tfrac{0.7625}{s+0.7265} \qquad \text{... 11.17}$$

Now replace s by *(z-1)/(z+1)*, which leads to,

$$H(z) = \tfrac{0.4201(1+z^{-1})}{1-0.1584z^{-1}} \qquad \text{... 11.18}$$

After performing an inverse Z-transform on this expression we arrive at a difference equation of the filter which can be implement (programmed) directly,

$$y(n) = 0.4208[x(n)+x(n-1)] + 0.1584y(n-1) \qquad \text{... 11.19}$$

We can now use DADiSP to simulate the behaviour of this filter. First create an impulse.

W1: gimpulse(300,.0066)

Now create the filter; its transfer function is shown in Figure 11.5.

W2: Filteq({0.4208, 0.4208}, {1, 0.1584}, w1)

You will observe from Figure 11.5 the transfer function is that of a low pass filter with a very gentle roll-off from 30 Hz is 6.5 dB/octave (which can be measured using the **Crosshair** at 30 Hz to 60 Hz).

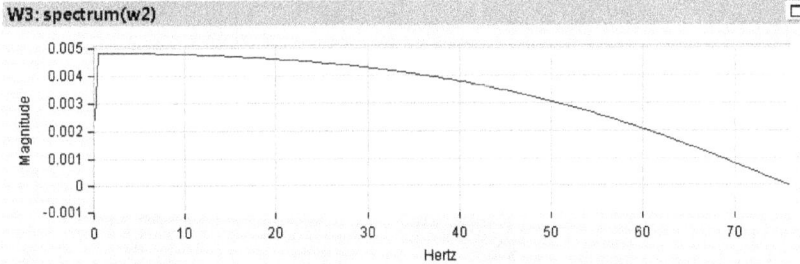

Figure 11.5: A single pole low pass filter derived from Eq:11.19

Referring back to Figure 11.1. Here are a few useful rules.

- The execution time of a filter is dependent upon the number of stages (coefficients) - it's customary to minimise the execution time of the filter.
- The first choice is the Butterworth filter (typically 6 dB/octave roll-off per order) as it has no ripple in either the pass band or stop band.
- For demanding specifications Butterworth filters require more stages and takes longer to execute. If this is unacceptable consider the other designs.
- If low execution time is critical and you require a very high performance filter your choice will be the multistage Elliptical filter.

11.3 Group Delay

If you are interested in the phase response you need to examine the *group delay* for the filter. This is defined as,

$$\frac{d \arg[H(\omega)]}{d\omega} \qquad \qquad ... 11.20$$

where *arg[H(ω)]* is the imaginary part of the transfer function *H(ω)*. This is the *gradient* of the phase spectrum which provides useful information with regards to the phase behaviour across the whole frequency range of a filter. An example of how severe the group delay is shown in Figure 11.6 for a Chebyshev filter. A you can see there is severe delay at the cut-off region of the filter. Any phase information in the input signal would be severely compromised if passed through this filter. Compare this with a symmetric FIR filter where the group delay is *flat* right across the spectrum. If you require a near zero group delay with a IIR filter you may wish to consider a Bessel filter design. There is a command in DADiSP for determining the group delay of a transfer function *H(z)*.

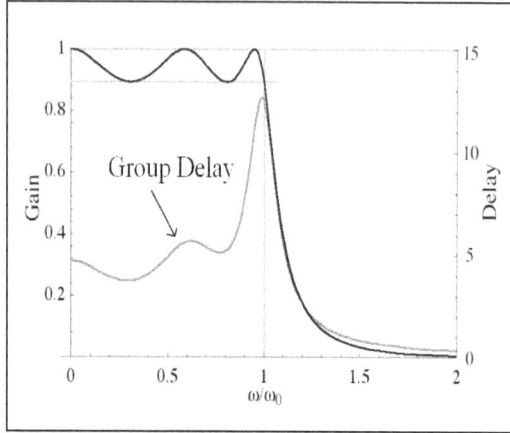

Figure 11.6: An example of group delay in a Chebyshev I filter

Consider the example of a transfer function,

$$H(z) = \frac{1.3 - 0.654z^{-1} + 1.43z^{-2}}{1 - 0.876z^{-1} + 0.9z^{-2}} \qquad \dots 11.21$$

The DADiSP command for calculating the group delay for this transfer function is,

W1: grpdelay({1.3, -0.654, 1.43}, {1, -0.876, 0.9}, 1024)

The result is shown in Figure 11.7. The group delay shown in Figure 11.7 is quite distinctive and the plot of the transfer function of Eq:11.21 is shown in Figure 11.8. To generate Figure 11.8, an impulse of 1024 data values was needed (using the *gimpulse* command) and the use of the *Filteq* command to find the impulse response for the Eq:11.21. Figure 11.8 clearly shows the presence of pole and a zero both of which have a profound effect upon the group delay as shown in Figure 11.7. The two peaks in the group delay correspond with the locations of the pole and zero.

W4: grpdelay({1.3, -0.654, 1.43}, {1, -0.876, 0.9}, 1024)

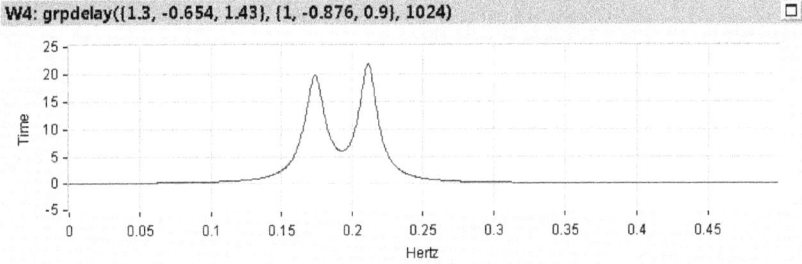

Figure 11.7: The group delay for the transfer function in Eq:11.21

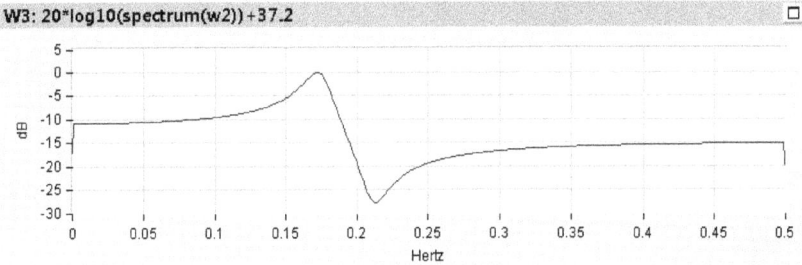

W3: 20*log10(spectrum(w2))+37.2

Figure 11.8: Transfer function of Eq:11.21

11.4 Commercial Filter Design Software

The general expression for a filter realised from biquadratic sections is,

$$H(z) = \prod_{i=1}^{K} \frac{b(i,0)+b(i,1)z^{-1}+b(i,2)z^{-2}}{1+a(i,1)z^{-1}+a(i,2)z^{-2}} \qquad ... 11.22$$

When using commercial software for determining the coefficients $\{a\}$ and $\{b\}$, they are usually produced for implementation on cascaded biquadratic sections as seen in Table 11.1. The coefficients in Table 11.1 are un-quantised; using these coefficients leads to a perfect simulation. The information in Table 11.1 is an example of an Elliptical IIR filter and each section is numbered (I) which contains three $\{b\}$ coefficients and two $\{a\}$ coefficients. For more demanding filters where the specifications are tighter, you will find more stages. In the following example there is an illustration how each stage of a IIR filter performs before passing into the next stage. In effect, using DADiSP to simulate and observe how each stage of this Elliptical filter performs. Note *each stage feeds into the next stage.* First generate in W1 an impulse with 300 samples and a sample separation which corresponds to $1/fs = 0.0005$s.

W1: gimpulse(300, 0.0005)

INFINITE IMPULSE RESPONSE (IIR)
ELLIPTICAL LOWPASS FILTER
UNQUANTIZED COEFFICIENTS
FILTER ORDER = 7

SAMPLING FREQUENCY = 2.000 KILOHERTZ

I	A(I,1)	A(1,2)	B(I,0)	B(I,1)	B(I,2)
1	-0.790103	0·000000	0·104948	0·104948	0·000000
2	-1.517223	0·714088	0·102450	-0·007817	0·102232
3	-1.421773	0·861895	0·420100	-0·399842	0·419864
4	-1.387447	0·962252	0·714929	-0·826743	0·714841

****CHARACTERISTICS OF DESIRED FILTER ***

	BAND 1	BAND 2
LOWER BAND EDGE	0·0000	·30000
UPPER BAND EDGE	·25000	1·00000
NOMINAL GAIN	1·0000	·00000
NOMINAL RIPPLE	·05600	·00100
MAXIMUM RIPPLE	·04910	·00071
RIPPLE IN DB	·41634	-63·00399

Table 11.1: IIR filter specifications and coefficients

Now create **Stage 1** of the filter - the response is shown in Figure 11.9.

W2: Filteq({0.104948, 0.104948, 0}, {1, -0.790103, 0.}, w1)

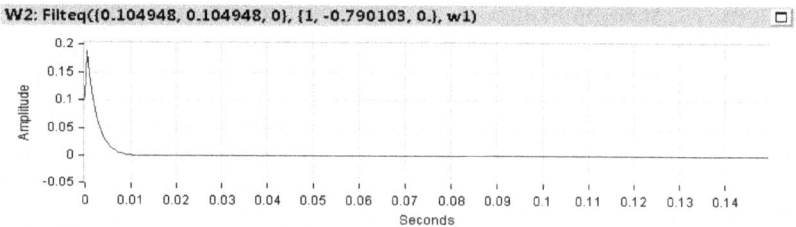

Figure 11.9: The impulse response to Stage 1 of the IIR filter

Create **Stage 2** of the filter which is shown in Figure 11.10.

W3: Filteq({0.10245, -0.007817, 0.102232}, {1, -1.517223, 0.714088}, w2)

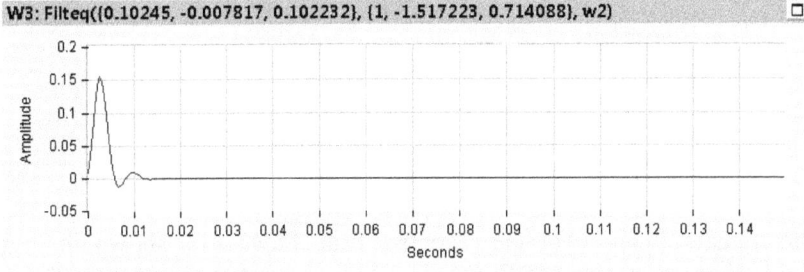

Figure 11.10: The impulse response of Stage 2 of the IIR filter

You will observe from Figure 11.10 the impulse response is taking on the type of waveform you would expect at this stage.

Create the **Stage 3** of the filter which is shown in Figure 11.11.

W4: Filteq({0.420100, -0.399842, 0.419864}, {1, -1.421773, 0.861895}, w3)

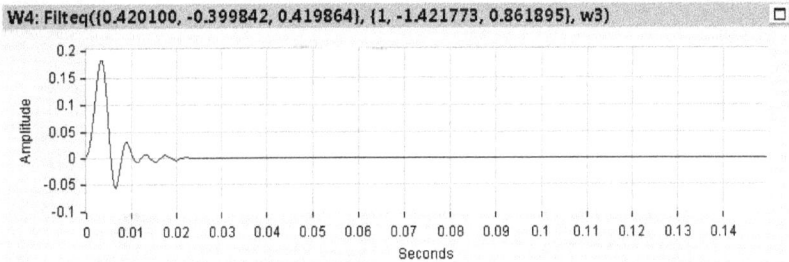

Figure 11.11: The impulse response to Stage 3 of the IIR filter

And finally create the **Stage 4** which is shown in Figure 11.12.

W5: Filteq({0.714929, -0.826743, 0.714841}, {1, -1.387447, 0.962252}, w4)

W5: Filteq({0.714929, -0.826743, 0.714841}, {1, -1.387447, 0.962252}, w4)

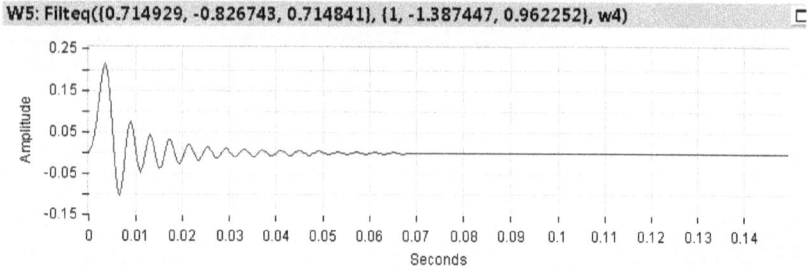

Figure 11.12: The impulse response to Stage 4 of the IIR filter

The transfer function of the response from Stage 4 is shown in Figure 11.13. You will observe in Figure 11.13 the characteristic ripple in the pass band of the Elliptical filter. Compare with Figure 11.1 which comprises four classes of IIR filter profiles. You will note from Figure 11.12 the impulse response of the Elliptical filter contains damped ringing. This is quite normal for a high performance filter which has poles close to the perimeter of the unit circle. To observe the ripple in the stop band a log scale is required on the *y*-axis as shown in Figure 11.14.

W6: spectrum (w5)

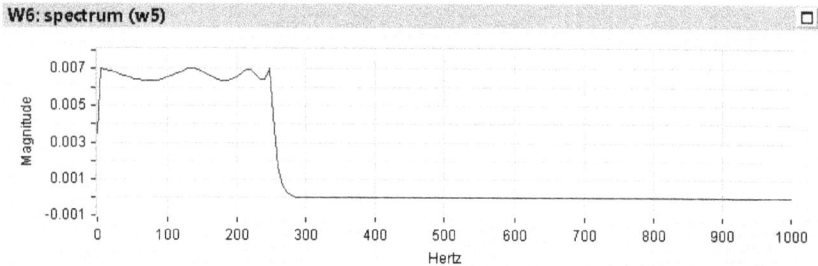

Figure 11.13: The spectrum of the filter specified in Table 11.1

The transfer function in Figure 11.14 has been scaled to give the pass band of 0 dB. As can be observed over a frequency range of 50 Hz (250Hz → 300Hz) the transmission has dropped -60dB. You will also observe the performance of the filter corresponds closely to the specification in Table 11.1. You may also wish to compare Figure 11.14 with Figure 11.1 which also shows the transfer function of an Elliptical filter with a dB scaling.

W6: 20*log10(spectrum (w5))+44

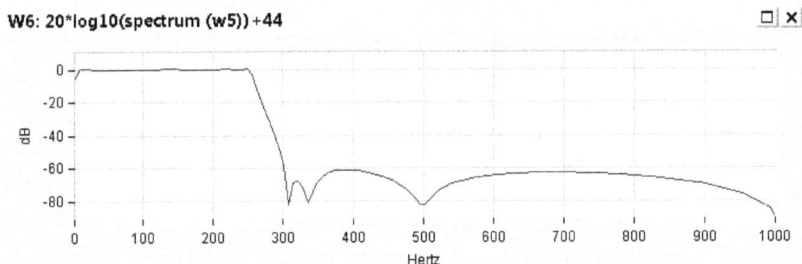

Figure 11.14: dB scale for the filter specified in Table 11.1

Using the IIR Design in DADiSP

When it comes to specifying IIR filters the facility in DADiSP is remarkably effective. To illustrate this feature we shall work through an example, the design of a high pass filter with the following specifications;

- Filter Design → Chebyshev 1
- Sampling Rate → 3,000Hz
- Cut-off Freq → 600 Hz
- Stopband Edge → 900 Hz
- Cascade, Bilinear Transform
- Pass/Stop Ripple (dB) → 0.5
- Stopband atten (dB) → 70.

Click on Filters and select IIR Filter Design, you will be presented by a dialogue box as shown in Figure 11.15. Complete the fields as shown in Figure 11.15 with the above specifications. After you click on OK, press the F7 key six times and table as shown in Figure 11.16 will appear. To observe the transfer function for this filter, click on Filters and select Filter Response. In the dialogue box which opens, select Magnitude, Cascade, Cascade Coefficients W1, Number of Samples 1024 and Destination W2. When you click on OK you will see the transfer function as shown in Figure 11.17. From Figure 11.17 you will observe the transfer function which has the characteristic ripple in the pass band where the stop band starts at 900Hz.

157

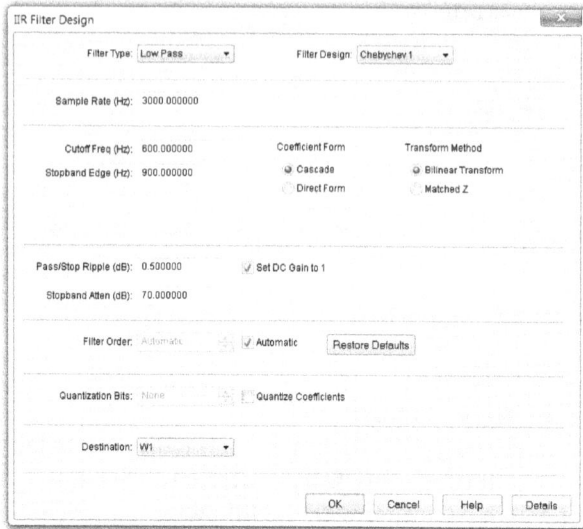

Figure 11.15: dialogue box for IIR filter design

Figure 11.16: Table with the coefficients for the IIR filter

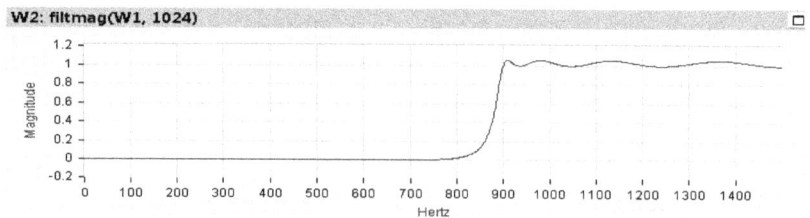

Figure 11.17: The Transfer function of the IIR filter

A plot of the features in the unit circle can be obtained by again selecting Filters→ **Other Functions**. In the dialogue box, select **Pole Zero Plot, Cascade, Cascade Coefficients** W1, **Destination:** W3. After clicking OK, you will see the unit circle as shown in Figure 11.18 which is characteristic of the pole locations for a Chebyshev filter. You are now encouraged to experiment with the IIR filter design in DADiSP, don't forget to keep a print of any interesting filters you design.

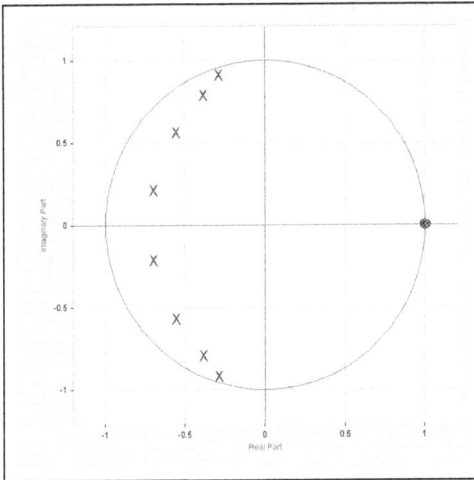

Figure 11.18: Poles in the unit circle

11.5 Stability and Quantisation Effects

IIR filters by their recursive nature can become unstable - too much feed back can destabilise any system. Considerable effort has been spent into understanding the causes of instability in IIR filters and digital systems in general. One such cause is *coefficient quantisation*. To gain an insight into the significance of quantisation consider the straight line shown in Figure 11.19.

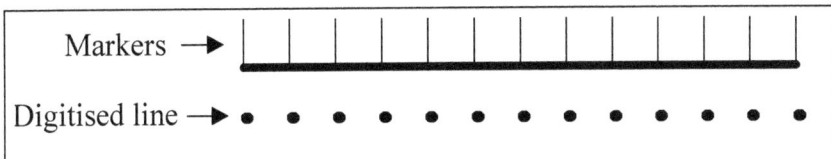

Figure 11.19: The effect of digitising a straight line

When considering the straight line with markers in Figure 11.19, you will observe continuity between each marker. In other words there is an infinite precision between each marker. By this we mean there is an infinite number of fractional values between each marker. Once you digitise (quantise) the line - converting it into a set of samples - the high precision has been lost. You have vacant gaps between the points. The precision is dependent upon the gaps between the samples - the smaller the gaps the better the precision. A similar process occurs when coefficients are quantised. Instead of being able to place a pole anywhere in the unit circle, they locate to the nearest quantised spot. If a pole in a IIR filter passes outside the unit circle the filter becomes unstable.

11.6 Sensitivity of Poles to Coefficient Quantisation

What is required is a method to ensure that when quantisation occurs the sensitivity to the pole positions is minimised. Given a transfer function,

$$H(z) = \frac{\sum\limits_{k=0}^{k=M} b_k z^{-k}}{1 + \sum\limits_{k=1}^{k=N} a_k z^{-k}} \qquad \text{... 11.23}$$

when the coefficients are quantised the transfer function will become modified,

$$\bar{H}(z) = \frac{\sum\limits_{k=0}^{k=M} \bar{b}_k z^{-k}}{1 + \sum\limits_{k=1}^{k=N} \bar{a}_k z^{-k}} \qquad \text{... 11.24}$$

where the coefficients now become,

$$\bar{a}_k = a_k + \delta a_k, \; k = 1, 1, 2, ... N$$
$$\bar{b}_k = b_k + \delta b_k, \; k = 0, 1, 2 ... M \qquad \text{... 11.25}$$

In these coefficients $|\delta a_k|$ and $|\delta b_k|$ represent the errors due to quantisation. In fact the quantisation process is thought of as adding noise to the system. Looking at Eq:11.23, the denominator is,

$$D(z) = 1 + \sum\limits_{k=1}^{k=N} a_k z^{-k} = \prod\limits_{k=1}^{n=N} (1 - p_k z^{-1}) \qquad \text{... 11.26}$$

When quantisation takes place the denominator $D(z)$ will change, Eq:11.26 therefore becomes,

$$\bar{D}(z) = 1 + \sum_{k=1}^{k=N} \bar{a}_k z^{-k} = \prod_{k=1}^{n=N}(1 - \bar{p}_k z^{-1}) \qquad \ldots 11.27$$

where $\bar{p}_k = p_k + \delta p_k$ is the new location of the kth pole as a result of the quantisation. Expressing the pole position error as,

$$\delta p_i = \sum_{k=1}^{k=N} \frac{\partial p_i}{\partial a_k} \delta a_k \qquad \ldots 11.28$$

Eq:11.28 is extensively used in general *sensitivity analysis* of systems and in this case the expression gives an estimate of the sensitivity of the pole positions as a result of quantising the coefficients $\{a\}$. After some algebra is can be shown the error in the pole position is given by,

$$\delta p_i = \sum_{k=1}^{k=N} \frac{p_i^{N-k}}{\prod_{\substack{l=1 \\ l \neq i}}^{l=N}(p_i - p_l)} \delta a_k \qquad \ldots 11.29$$

You will observe the error in the pole position values (δp_i) is at a minimum when the product in the denominator in Eq:11.29 is maximised. $\prod(p_i - p_l)$ represents the product of all the pole separations. Therefore by maximising the pole separation you minimise the effect of quantising the coefficients on the transfer function $H(z)$. What can so easily happen in a filter, where the poles are tightly clustered, the quantisation of $\{a\}$ will have a significant effect on $H(z)$. Very often some of these poles, which lie very close to the perimeter of the unit circle, when quantising the $\{a\}$ coefficients, one or more of the poles could be forced outside the unit circle. When this happens the filter becomes unstable and therefore unusable. To maximise the value of $\prod(p_i - p_l)$, filters are realised as double pole sections. Typically a series of biquadratic sections as shown in Figure 10.12. Just to reinforce the point, Table 11.1 shows the coefficients already displayed for a biquadratic structured filter.

Two Pole Example
To gain a further insight into the possible positioning of poles in the unit circle, consider the IIR filter,

$$H(z) = \frac{1}{1 - 2r\cos(\theta) + r^2 z^{-2}} = \frac{1}{(1 - re^{j\theta}z^{-1})(1 - re^{-j\theta}z^{-1})} \qquad \ldots 11.30$$

The coefficients are $a_1 = 2r\cos(\theta)$ and $a_2 = -r^2$ and when viewing Figure 11.20 the quantised pole locations are formed by the intersection of the vertical lines defined by the quantised value of $2r\cos(\theta)$ concentric rings whose radii are defined by the quantised value of the square root of r^2. The less the degree of quantisation the more rings you have in the unit circle and the closer the possible pole positions. You will observe that the density of pole positions increase as you approach the perimeter of the unit circle. However the number of positions within the unit circle are still limited and its quite possible for poles to locate outside the circle which leads to the filter being unusable.

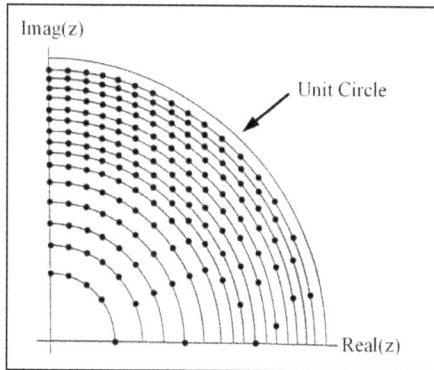

Figure 11.20: The location of possible pole locations after coefficient quantisation has taken place

When implementing IIR filters on digital signal processors the primary choice is a *floating point* processor such as the TMS320C6701 from Texas Instruments. To understand why floating point numbers are desirable here is a brief overview.

Floating Point and Fixed Point
For an n-bit binary number instead of representing a fraction with n places after the binary point, a number of the bits are used to represent an exponent.

S	•1	0	1	1	0	0	0	1	0	1	1	1	0	1	1

This is a 16-bit number *fixed point* where all the bits apart from the first bit are used to represent a fractional number; • shows the location of the *binary point* (equivalent to the decimal point). The first bit represents the sign of the number; 0 for a positive number and 1 for a negative number.

S	•1	0	1	1	0	0	0	1	0	1	s	1	0	1	1

In a floating point binary number, some of the bits are used to represent an exponent - marked in bold. A floating point binary number is therefore expressed as $(-1)^s \cdot c \times 2^q$ where c is the mantissa (the fractional part), and q the exponent and s the sign of the exponent. In the above example the floating point number would be,

$$(-1)^S \cdot 1011000101 \times 2^{s1011}$$

The range of numbers which can be represented by floating point formats is far greater than for fixed point. The internal architecture of a FP DSP is more complex as the exponent has to be treated differently from that of the mantissa. When multiplying two floating point numbers, the mantissa are multiplied and the exponents are added. The international standard for representing floating point numbers is known as the IEEE-754 which has provision for different levels of precision. In particular, *single precision* (32-bit), *double precision* (64-bit) and *extended precision* (80-bit). When writing code in C++ a programmer defines at the start of program the precision of the variables (for example #*double x, y*). Highly recursive code would perform better in high precision although they take longer to execute.

However it's not always cost effective to use a floating point processor in which case special measures have to be taken when implementing IIR filters on fixed point DSPs.

What you have gained from this Chapter

1. An understanding of the nature of recursion and feedback.
2. An introduction to impulse invariant design method and the problem of aliasing.
3. An insight into the bilinear transform and the process of mapping functions.
4. Implementing IIR filters on DADiSP
5. The concept of group delay.
6. A brief look at commercial software for IIR design.
7. Designing IIR filters using DADiSP and viewing the transfer function of the filters.

8. An understanding of the stability of IIR filters and the effect of quantisation.
9. A look at the pole positions for a two pole filter.
10. A brief insight into fixed and floating point numbers.

The DADiSP skills you have acquired from this Chapter

1. *fimpulse* - method for generating an impulse.
2. *filteq* - creating IIR filter from a set of coefficients.
3. *spectrum* - obtaining a spectrum of a signal
4. *grdelay* - for obtaining the group delay of a IIR filter
5. *Cheby1* - designing a IIR filter
6. *Filtmag* - obtaining the transfer function of a filter
7. *zplane* - generating a unit circle showing the pole positions

DADiSP Extra

To generate a spiral trace, enter the following,

W1: gsin(200,.001,40)*gexp(200,.001,-10)

W2: gcos(200,.001,40)*gexp(200,.001,-10)

In W3, enter XY(W1,W2) and you should see a spiral trace as shown in Figure 11.21.

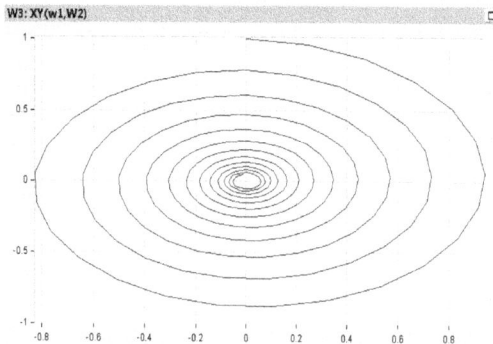

Figure 11.21: An example of a spiral trace

12. Spectral Analysis

Spectral analysis accounts for a great deal of digital signal processing activity and the principal tool for performing this analysis is the Fourier Transform (FT). The Fourier Transform, performed over a time T, is defined as,

$$X(\omega) = \frac{1}{T} \int_0^T x(t)\, e^{-j\omega t} dt \qquad \ldots 12.1$$

$X(\omega)$ represents the frequency contents within the signal $x(t)$. In effect it performs like a glass prism when white light is passes through it - the light is dispersed into seven colours. Although the Fourier Transform is the most widely used tool for deriving the frequency content of a signal, it is by no means the only one, but it is the most understood. So how does the Fourier Transform work? To answer this question we need to introduce the *orthogonal* function.

12.1 Orthogonal Functions

The word orthogonal means at right angles ⊥. The mathematical definition of an orthogonal function relates to,

$$\langle f, g \rangle = \int f^*(x)\, g(x)dx = 0 \quad if \ f \neq g \qquad \ldots 12.2$$

This represents the *inner product* which is only finite if $f = g$ otherwise it is zero. The following are orthogonal functions,

$$\frac{1}{\pi} \int_{-\pi}^{\pi} \sin(nx)\sin(mx)dx = \begin{array}{l} 0 \ \ when \ \ n \neq m \\ 1 \ \ when \ \ n = m \end{array} \qquad \ldots 12.3$$

$$\frac{1}{\pi} \int_{-\pi}^{\pi} \cos(nx)\cos(mx)dx = \begin{array}{l} 0 \ \ when \ \ n \neq m \\ 1 \ \ when \ \ n = m \end{array} \qquad \ldots 12.4$$

$$\int_{-\pi}^{\pi} \sin(nx)\cos(mx)dx = 0 \qquad \ldots 12.5$$

You will observe from Eq:12.3 and Eq:12.4 these integrals only produce finite results when $n = m$, otherwise they are zero. In Eq:12.5 the integral is always zero. We now relate these orthogonal functions to the expanded Fourier Transform,

$$X(\omega) = \int x(t)\cos(\omega t)dt - j \int x(t)\sin(\omega t)dt \qquad \text{... 12.6}$$

If we consider a single frequency ω_o, then

$$X(\omega_o) = \int x(t)\cos(\omega_o t)dt - j \int x(t)\sin(\omega_o t)dt \qquad \text{... 12.7}$$

x(t) is a signal which may contain several sine and cosine waves. If we now compare Eq:12.7 with Eq:12.3 and Eq:12.4, we will observe,

1. the first integral in Eq:12.7 is only finite if *x(t)* contains cos*(ω,t)* and
2. the second integral is only finite if *x(t)* contains sin*(ω,t)*.

This means that $|X(\omega_o)|$ is only positive if the frequency ω_o is present in *x(t)*, otherwise it is zero. The Fourier Transform performs this orthogonality test on every frequency in *x(t)*. It therefore isolates each frequency contained in *x(t)* which is how it identifies all the frequencies present in *x(t)*.

12.2 Interpreting the Fourier Transform

Remember *X(ω)* is complex and contains two components of spectral information $\text{Re}[X(\omega)]$ and $\text{Im}[X(\omega)]$; $|X(\omega)|$ is the magnitude of the spectral contents and is defined as,

$$|X(\omega)| = \sqrt{\text{Re}[X(\omega)]^2 + \text{Im}[X(\omega)]^2} \qquad \text{... 12.8}$$

And the phase spectrum is,

$$\theta(\omega) = \tan^{-1}\left[\frac{\text{Im}[X(\omega)]}{\text{Re}[X(\omega)]} \right] \qquad \text{... 12.9}$$

Note the limits on the value of the phase $-\frac{\pi}{2} < \theta(\omega) < \frac{\pi}{2}$. Very often when you see the phase spectrum of signals you will often see the *fly-back* in the spectrum. As an example consider the FIR filter,

$$y(n) = x(n) + x(n-3) + x(n-4) \qquad \text{... 12.10}$$

Generate an impulse in W1 and then use the Filteq command

W1: gimpulse(300,.01)

W2: Filteq({1,0,0,-1,-1}, {1}, w1)

W3: phasespec(w2)

The phase spectrum of the response function in W3 is shown in Figure 12.1.

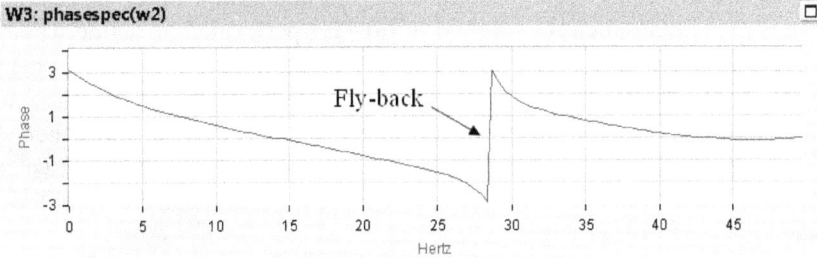

Figure 12.1: Phase spectrum of FIR filter - Eq:12.10

You will observe in Figure 12.1 the fly-back takes place in the phase spectrum. What happens in reality; in this example the curve continues downward however the fly-back is needed owing to the limits on $\theta(\omega)$ - the calculation does not go beyond $\pm\pi$ so it flies-back. The spectrum of the impulse function is shown in Figure 12.2.

W4: spectrum(w2)

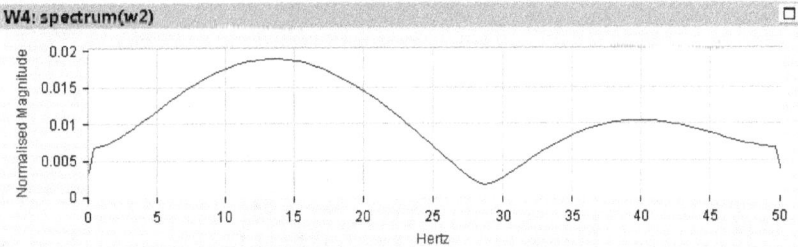

Figure 12.2: Transfer function for the filter defined by Eq:12.10

In general it is the magnitude spectrum which is usually more useful and is usually the default setting for commercial spectrum analysers. An example of such a spectrum analyser is the *Agilent 35670A* which is featured in Chapter 14.

12.3 Spectral Resolution

Spectral resolution is the ability of any process to resolve spectral lines and this is dependent upon several factors which shall be discussed in this section. Eq:12.1 shows the Fourier Transform with the integration being performed over a finite time T as opposed to an infinite integration time. What are the consequences of limiting the integration time? In Chapter **8** we showed that when the integral in Eq:12.1 performed on a sine wave the result is a sinc function in the frequency domain. Not only do you get the central lobe but also side lobes where there has been leakage from the main lobe. What happens if there are two spectral lines very close together? We can use DADiSP to see the outcome. Generate a signal with two sine waves, one at 10Hz and the second at 11Hz both of the same amplitude as seen in Figure 12.3

W1: gsin(300,.01,10)+gsin(300,.01,11)

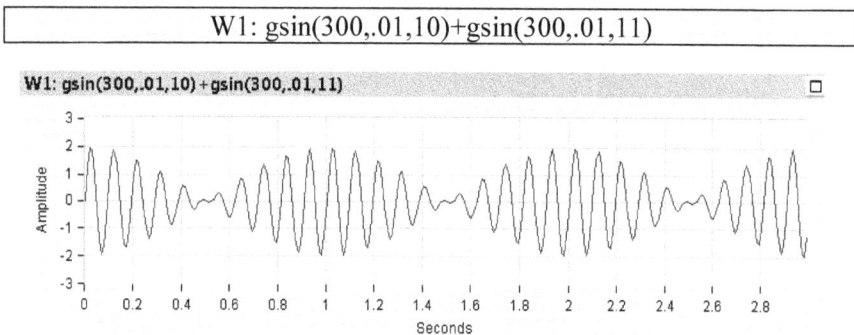

Figure 12.3: Two sine waves with very close frequencies

The spectrum of this waveform is shown in Figure 12.4.

W2: spectrum(w1)

As can be seen in Figure 12.4 both spectral lines have been resolved. Now reduce the integration time by reducing the number of samples to 100. The new waveform is shown in Figure 12.5. The corresponding spectrum of Figure 12.5 is shown in Figure 12.6. You will observe from Figure 12.6 there is only one broad spectral line - the individual lines have failed to be resolved. Spectral resolution is therefore concerned with ensuring all spectral lines are resolved however close in frequency they may be.

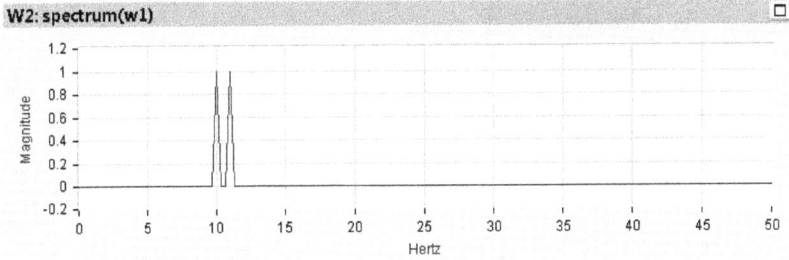

Figure 12.4: Spectrum of the waveform in Figure 12.3

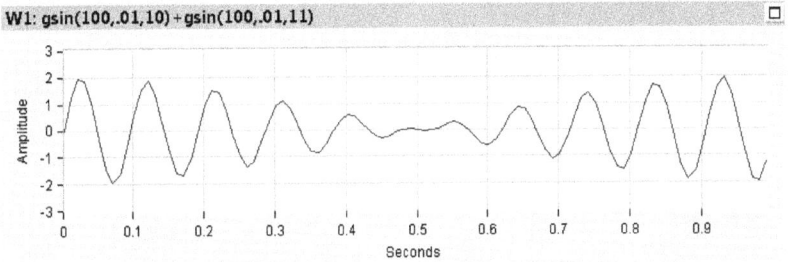

Figure 12.5: The waveform in Figure 12.3 with fewer samples

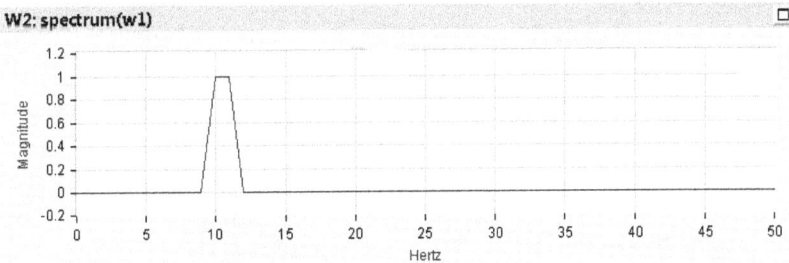

Figure 12.6: Spectrum of the waveform in Figure 12.5

Considering the waveform in Figure 12.5, if the number of data points is increased to 260, the resulting spectrum is shown in Figure 12.7. In Figure 12.7 the spectral lines are regarded as partially resolved and in Chapter 13 a method is presented for determining how many data samples are required to resolve two closely spaced spectral lines.

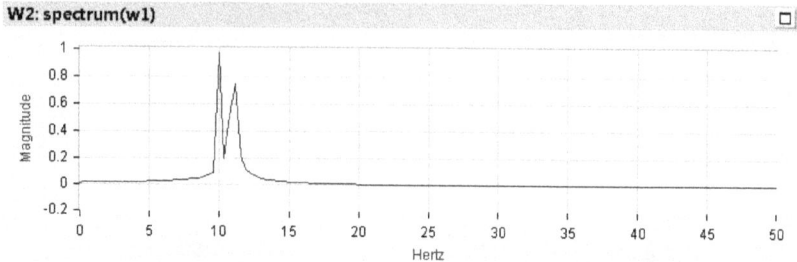

Figure 12.7: Partially resolved spectral lines

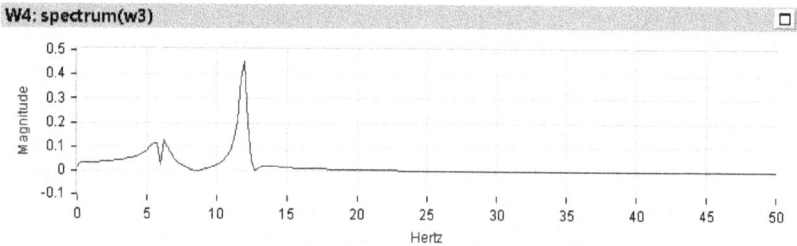

Figure 12.8: Spectrum of transient waveform in Figure 2.22

12.4 Transient Signals

In Chapter 2 you were introduced to a variety of signals including transient signals. As an exercise, create a transient signal as prescribed in Chapter 2 and obtain a spectrum of it. You should see a spectrum similar to that in Figure 12.8. This spectrum contains a number of interesting features. In the first instant the major peak has frequency of 12 Hz as expected from Figure 2.19. There should also be a spectral peak at 6 Hz when in fact there is a noticeable dip in the spectrum (use the Cursor → Crosshairs to confirm this result).

Task: Provide an explanation why there is no spectral line at 6 Hz in Figure12.8.

Make adjustments to the signal in W1 and observe the spectrum obtained in W4.

12.5 Signals in Noise

The signals analysed so far in this chapter have been free of noise, in reality many signals are contaminated by noise and whatever spectral measurements are made on a signal there will be a noise component. There are however methods for increasing the accuracy of a spectral measurement by considering the *power spectral density*. First it is necessary to consider the energy in a waveform, this is expressed as,

$$E_x = \int\limits_{-\infty}^{+\infty} |x(t)|^2 dt \qquad \text{... 12.11}$$

In view of the fact that *x(t)* could be complex, the energy content becomes,

$$E_x = \int\limits_{-\infty}^{+\infty} x(t)x^*(t)dt \qquad \text{... 12.12}$$

By introducing the inverse Fourier Transform into this expression,

$$E_x = \int\limits_{-\infty}^{+\infty} x(t)\left[\int\limits_{-\infty}^{\infty} X^*(\omega)e^{-j\omega t}d\omega \right]dt \qquad \text{... 12.13}$$

Alternatively,

$$E_x = \int\limits_{-\infty}^{+\infty} X^*(\omega)d\omega\left[\int\limits_{-\infty}^{\infty} x(t)e^{-j\omega t}dt \right] \qquad \text{... 12.14}$$

Which leaves,

$$E_x = \int\limits_{-\infty}^{+\infty} |X(\omega)|^2 d\omega \qquad \text{... 12.15}$$

But according to Eq:12.12,

$$E_x = \int\limits_{-\infty}^{+\infty} |x(t)|^2 dt = \int\limits_{-\infty}^{+\infty} |X(\omega)|^2 d\omega \qquad \text{... 12.16}$$

This is referred to as *Parseval's Theorem* which states the energy in the time domain is the same as the energy in the frequency domain. Now introduce,

$$S_{xx}(\omega) = |X(\omega)|^2 \qquad \text{... 12.17}$$

which in effect represents the distribution of energy across the frequency spectrum generated from the signal *x(t)* and is referred to as the *energy spectral density*. $S_{xx}(\omega)$ is in fact symmetrical, therefore,

$$S_{xx}(-\omega) = S_{xx}(\omega)$$ ... 12.18

The autocorrelation function for a signal *x(t)* is given by,

$$R_{xx}(\tau) = \int x(t)\, x(t - \tau)dt$$ ... 12.19

The Fourier Transform of $R_{xx}(\tau)$ is given by,

$$P_{xx}(\omega) = \int R_{xx}(\tau) \cos(\omega\tau)d\tau$$ ... 12.20

$P_{xx}(\omega)$ is referred to as the *power spectral density* (PSD) and since it is derived from the autocorrelation function $R_{xx}(\tau)$ it contains no phase information. To illustrate the advantage of using the PSD, create a sine wave with added noise in W1, its dB spectrum in W2 and its dB PSD in W3,

W1: gsin(300,.001,36)+gnormal(300,.001,0,.5)

W2: 20*log10(spectrum(w1)

W3: 20*log10(spd(w1))

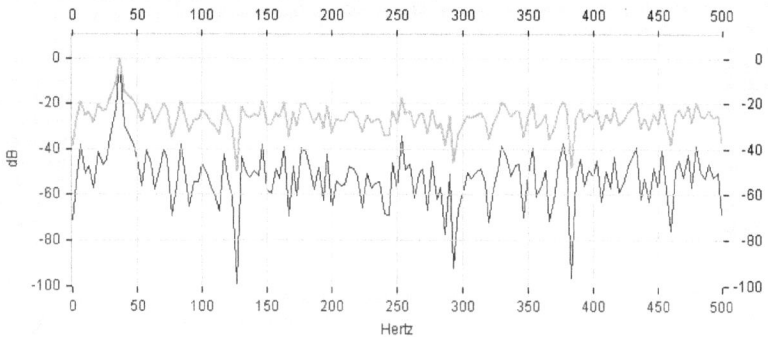

Figure 12.9: A spectrum (upper trace) and a PSD (lower trace) displayed together

Figure 12.9 shows the overlay of the spectrum of the waveform (upper trace) and its power spectral density calculated from the Fourier Transform of its autocorrelation function (lower trace). You will observe the significant reduction in the noise when using the PSD; *mean(W2)* = -26.2 dB, *mean(w3)* = -52.3 dB. The PSD is a measurement to consider when dealing with signals which are known to have a significant noise content. As an exercise, make adjustments to W1 by adding another frequency component of lower amplitude and determine how well each spectral process performs - don't forget to keep a printed copy of your results as the knowledge you gain from this exercise may become useful later in your career as an electronic engineer or a mechanical engineer.

12.6 The Spectrogram

An alternative method for analysing spectra is the spectrogram which is a highly visual technique by making use of colour. A signal is sliced up into segments and a Fourier Transform is performed on each segment. In the spectrograph display, the *x*-axis is time, the *y*-axis is frequency and the colour represent the magnitude of the spectral content. DADiSP has a spectrogram command, but first it is necessary to import a signal. This can be achieved using the WAV command from the icon command line. To generate the spectrogram use,

W3: Specgram(W2, 512, 256, 512, 2)

In this instruction, the input file is in W2, the segment length is 512 samples with a sample overlap of 256 samples and a 512 point FFT is used (see Chapter 15). The image shown in Figure 12.10 shows a spectrogram and waveform of the word *hello* which has been adjusted to limit frequency range to 3 kHz (as opposed to 20 kHz). Use has been made of *Rainbow* from the Color Shading. It is customary for spectrograms to have have a high colour content. Making effective use of the spectrogram does take a amount of practise and it is particularly useful for the analysis of speech patterns for monitoring the movement of *formants*. These are the major frequency components in a person's voice which change as words are generated. Traditional speech recognition has made extensive use of spectrograms as it provides an readily accessible display showing frequency against time which is often required for dynamic signals.

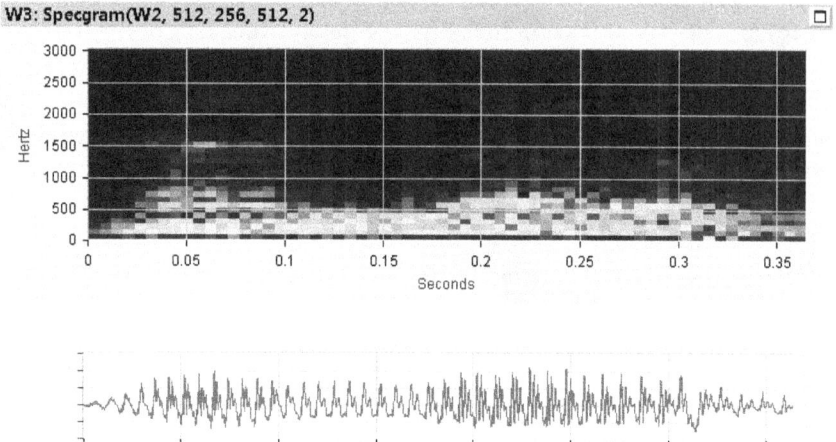

W3: Specgram(W2, 512, 256, 512, 2)

Figure 12.10: An example of a spectrogram of a waveform

What you have gained from this Chapter

1. An understanding of the Fourier Transform (FT) for obtaining spectral information from a signal.
2. The role played by orthogonal functions in understanding how the FT works.
3. Knowledge of the magnitude and phase spectra derived from the FT.
4. Understanding the nature of spectral resolution.
5. Understanding how to obtain spectral information from signals which have a high noise content.
6. An insight into the nature of a transient signal.
7. Parseval's Theorem and when to use the power spectral density measurement.
8. The spectrogram and its application in processing speech.

The DADiSP skills you have acquired from this Chapter

1. *gimpulse* - generate a impulse for testing systems.
2. *filteq* - generate the impulse response of filter.
3. *phasespec* - obtaining a phase spectrum of a waveform
4. *spectrum* - obtaining a spectrum of a signal
5. *gsin* - generate a sine wave

6. *gnormal* - generate a random signal where the samples have a normal distribution.

7. *20*log10(spectrum)* - to obtain a spectrum with a dB scale on the *y*-axis.

8. *psd* - to obtain the power spectral density of a signal.

9. *specgram* - to obtain a spectrogram of a signal.

10. *Cursor→ Crosshairs* - feature for measuring magnitude and frequency.

DADiSP Extra

Building on the theme of visualisation, in DADiSP you can construct radial sine wave patterns. In W1, click on f_x → **Generate Data** → **Z=F(X,Y)**. When the dialogue box opens, enter ther following,

Z=F(X,Y): → sin(4*r)   X Lower: → –4.0   X Upper: → 4.0
X Increment: → 0.1   Y Lower: → –4.0   Y Upper: → 4.0
Y Increment: → 0.1   Destination: → W4

When you click on OK you will see the following command in W1,

W1: sin(4*r);setplottype(4);setplotstyle(0)

Make adjustments in *Properties* for **Rainbow** from **Color Shading** and you should see the result as shown in Figure 12.11.

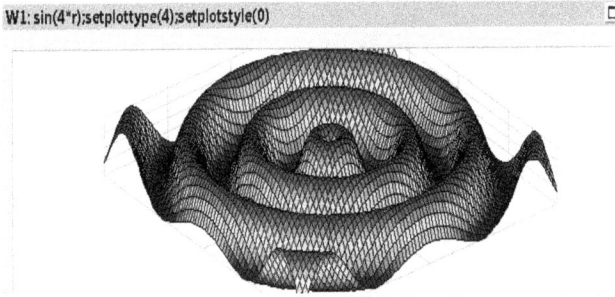

Figure 12.11: A radial sine wave

13. The Discrete Fourier Transform

You are now aware the Fourier Transform is one of the principle tools for deriving frequency information from a signal. This is expressed by,

$$X(f) = \frac{1}{T} \int_0^T x(n) \, e^{-j2\pi ft} dt \qquad \qquad ...13.1$$

The digital version is referred to as the *Discrete Fourier Transform* (DFT) and to gain an insight into it operation we need to consider the digitisation process. Figure 13.1 shows a digitised signal $x(t)$ where the samples are $x(n)$.

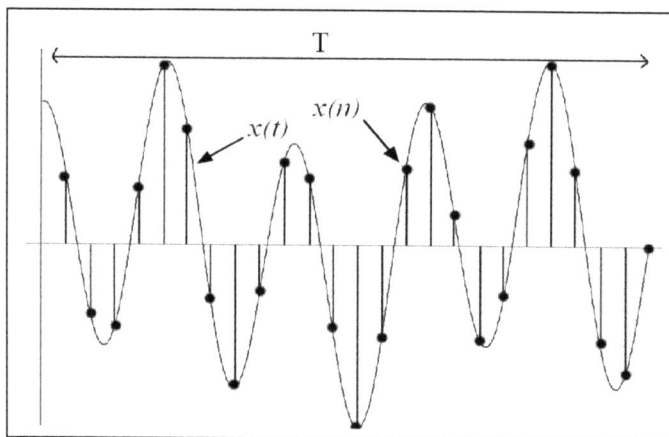

Figure 13.1: A digitised signal where the sample separation is Δ seconds

The digitised samples $x(n)$ is related to $x(t)$ by the expression

$$x(n) = x(t)|_{t=n\Delta} \qquad \qquad ...13.2$$

which indicates that samples are taken at intervals of Δ from $x(t)$. In Figure 13.1 you will observe the duration of the signal is T seconds and the interval between the samples is Δ. The FT is performed over the period T and if the number of samples is N then $T = N\Delta$. In Eq:13.1, the following analogue features are replaced by digital equivalents,

---

$\displaystyle\int_0^T \rightarrow \sum_{n=0}^{n=N-1}$ The integral is replaced by a summation where the limits reflect the duration of the integration.

$dt \rightarrow \Delta$ dt is replaced by Δ.

$f \rightarrow \frac{k}{N}$ f is replaced by k/N.

$t \rightarrow n$ t is replaced by n.

$x(t) \rightarrow x(n)$ $x(t)$ is replaced by $x(n)$.

$X(f) \rightarrow X(k)$ $F(f)$ is replaced by $F(k)$

$T \rightarrow N\Delta$ T is replaced by $N\,\Delta$

The DFT therefore becomes,

$$X(k) = \frac{1}{N} \sum_{n=0}^{n=N-1} x(n)\, e^{-j\frac{2\pi nk}{N}} \qquad \ldots 13.3$$

You will note that $X(k)$ is complex and contains both magnitude and phase spectral information.

13.1 Concept of the Frequency Bin

It is instructive to think of $X(k)$ comprising a set of *frequency bins* which are labelled by their k index as shown in Figure 13.2. If there are N data points in the DFT, there are $N/2$ *spectral bins* (the other $N/2$ spectral bins contain the same information as the first half).

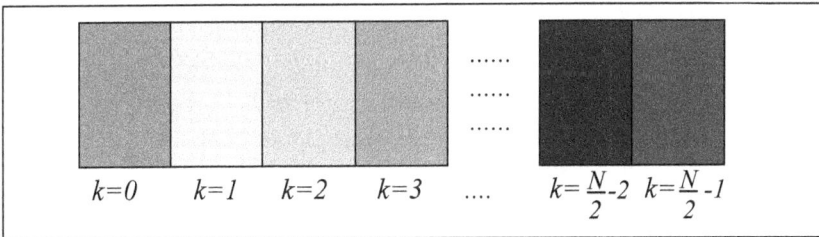

Figure 13.2: Frequency bins created by the DFT

An obvious question to ask is how wide is a spectral bin? The spectral range $f \in [0, fs/2]$ so the frequency bins are spread evenly over this frequency range. For example if the sampling frequency is 20kHz, the spectral range is from 0 to 10kHz. If there were 400 (N) data points in the sampled batch of data $\{x(n)\}$, there will be 200 frequency bins. The spectral width of each bin would be (10,000/200) Hz = 50Hz; the resolving power would therefore be 50Hz. You may well ask what happens to frequencies lower than 50Hz? Well they would all end up in the *first* frequency bin where $k = 0$. In effect, it will not be able to resolve frequencies lower than 50Hz or frequencies which are closer together than 50 Hz. To increase the resolution an increase in the number of samples in the sample batch is needed - thereby increasing the number of frequency bins spread between 0 and 10 kHz.

Example 13.1: If the time taken to collect a batch of samples from an analogue to digital converter is 4 seconds, and the sampling frequency is 300Hz, how many samples will be collected and what is the spectral width of each frequency bin?

Solution: In one second 300 samples will be collected, therefore over a period of 4 seconds, 1,200 samples will be collected. The number of frequency bins is $1,200/2 = 600$. If the spectral range is 150Hz and there are 600 bins, the width of each bin is,

$$\frac{150}{600} = 0.25 \text{ Hz.}$$

13.2 Computation of the DFT

You will notice from Eq:13.3 the presence of the complex exponential,

$$e^{-j\frac{2\pi nk}{N}} = \cos(\tfrac{2\pi nk}{N}) - j\sin(\tfrac{2\pi nk}{N}) \qquad \ldots 13.4$$

This contains sine and cosine terms which have to be calculated. There are number of ways how this task may be performed. In a number of Digital Signal Processors (DSPs) there are *look-up tables* in memory which hold a whole cycle of sine wave samples. These work on the principle that x acts as the address of the table entry and *sin(x)* is the contents of that address. If a look-up table is not available a calculation is required. Many functions can be approximated by using a polynomial series and in the case of the sine and cosine possible series are,

$$\sin(x) = x + b_3 x^3 + b_5 x^5 + b_7 x^7 + b_9 x^9 \qquad \ldots 13.5$$

$$\cos(x) = z + c_3 x^3 + c_5 x^5 + c_7 x^7 + c_9 x^9 \qquad \ldots 13.6$$

where $z = \frac{\pi}{2} - |x|$ and the coefficients are,

$$b_3 = -0.1666665668 \qquad b_5 = 0.0083302513$$
$$b_7 = -0.0001908074182 \qquad b_9 = 0.0000026019030$$
$$c_3 = -0.1666665668 \qquad c_5 = 0.008333025139$$
$$c_7 = -0.00019807418 \qquad c_9 = 0.000002601903$$

It is customary to express a series in the following format,

$$\sin(x) = x + b_3 x^3 + b_5 x^5 + b_7 x^7 + b_9 x^9$$
$$= x(1 + x^2(b_3 + x^2(b_5 + x^2(b_7 + b_9 x^2)))) \qquad ...13.7$$

You will observe the calculation is now a sequence of multiply and adds which ensures maximum precision. The calculation would begin with $b^9 x^2$ and progress towards the left. Performing Eq:13.3 with a data batch of N = 1000 data points requires quite a few calculations especially if you need to use the above sine and cosine expansions. If this series is implemented on a 24-bit DSP the coefficients in the series would be scaled (multiplied by $2^{23}-1$) and then be converted into hexadecimal numbers. To obtain the magnitude of the spectrum, the following calculation is needed,

$$|X(k)| = \sqrt{\text{Re}[X(k)]^2 + \text{Im}[X(k)]^2} \qquad ...13.8$$

Calculating a Square Root

Performing a square is not too difficult (multiplying two numbers together), however performing a square root is more difficult. One way of performing this calculation is to use the *Newton Raphson Algorithm* (NRA). It states that a real root of a function $f(y)$ can be calculated from an iterative process,

$$y_{n+1} = y_n - \frac{f(y_n)}{\left(\frac{df(y_n)}{dy_n}\right)} \qquad ...13.9$$

If $y = \sqrt{x}$, x is known and y is unknown. To use the NRA let,

$$f(y) = y^2 - x \qquad ...13.10$$

We therefore search for a value of y so that $f(y) = 0$ which means we have found the answer since $y = \sqrt{x}$. The NRA becomes,

$$y_{n+1} = y_n - \frac{y_n^2 - x}{2y_n} = 0.5(y_n + \tfrac{x}{y_n}) \qquad ...13.11$$

Since DSPs use fractional arithmetic then all values of $x < 1$. In the iterative procedure in Eq:13.11 a *seed value* is required, for example $y_0 = 0.5$. We can work through an example, to calculate $\sqrt{0.12}$, x is therefore 0.12. The value from a *Casio* pocket calculator is,

$$\sqrt{0.12} = 0.346410162.$$

Now start the iteration from Eq:13.11 with a seed value of 0.5,

$$y_1 = 0.5\left(0.5 + \frac{0.12}{0.5}\right) = 0.37$$

$$y_2 = 0.5\left(0.37 + \frac{0.12}{0.37}\right) = 0.347162162$$

$$y_3 = 0.5\left(0.347162162 + \frac{0.12}{0.347162162}\right) = 0.346410976$$

$$y_4 = 0.5\left(0.346410976 + \frac{0.12}{0.346410976}\right) = 0.346410162$$

After four iterations the NRA has yielded the same answer as a pocket calculator. Although this method can be used in principle you will observe it requires the use of a division in each iteration which in itself can be problematic. However many DSPs have an assembly language division instruction so this operation should not present too much of a problem. If the need arises to calculate the phase spectrum, the following series can be used to dteremine *atan(x)*,

$$\tan^{-1}(x) = \tfrac{\pi}{4} + c_1 y + c_3 y^3 + \ldots c_9 y^9 \qquad \ldots 13.12$$

where

$$y = \tfrac{x-1}{x+1} \qquad \ldots 13.13$$

The coefficients for the series in Eq:13.12 are,

$c_1 = 0.999866$	$c_3 = -0.3302995$
$c_5 = 0.180141$	$c_7 = -0.085133$
$c_9 = 0.0208351$	$\pi/4 = 0.785398$

You will observe that to use this series you need to apply Eq:13.13 to every value of x before Eq:13.12 can be used. To make effective use of these coeffients, the method laid down in Eq:13.7 needs to be used - a sequence of interlaced multiply and adds.

13.3 The DFT in Matrix Format

Very often the complex exponential in Eq:13.3 is expressed as,

$$e^{-j\frac{2\pi nk}{N}} = W_N^{nk} \qquad \dots 13.14$$

The DFT can therefore be expressed as,

$$X(k) = \frac{1}{N} \sum_{n=0}^{n=N-1} x(t)\, W_N^{nk} \qquad \dots 13.15$$

If we consider the DFT for 4 data values, then expanding Eq:13.15,

$$X(0) = \frac{1}{4}[x(0) + x(1) + x(2) + x(3)]$$

$$X(1) = \frac{1}{4}[x(0) + x(1)W_4^1 + x(2)W_4^2 + x(3)W_4^3]$$

$$X(2) = \frac{1}{4}[x(0) + x(1)W_4^2 + x(2)W_4^4 + x(3)W_4^6] \qquad \dots 13.16$$

$$X(3) = \frac{1}{4}[x(0) + x(1)W_4^3 + x(2)W_4^6 + x(3)W_4^9]$$

Eq:13.16 can now be represented in matrix format,

$$\begin{pmatrix} X(0) \\ X(1) \\ X(2) \\ X(3) \end{pmatrix} = \frac{1}{4} \begin{pmatrix} 1 & 1 & 1 & 1 \\ 1 & W_4^1 & W_4^2 & W_4^3 \\ 1 & W_4^2 & W_4^4 & W_4^6 \\ 1 & W_4^3 & W_4^6 & W_4^9 \end{pmatrix} \begin{pmatrix} x(0) \\ x(1) \\ x(2) \\ x(3) \end{pmatrix} \qquad \dots 13.17$$

An alternative way of expressing Eq:13.17 is,

$$X = \frac{1}{4} W_4 \cdot x \qquad \dots 13.18$$

where X is the column vector holding the spectral data, W_4 the *twiddle* matrix and x is the discrete time domain data vector. It is constructive to think of the matrix W_4 operating on the data vector x to provide the spectra data X. In principle the expression could be expanded to represent any size array of data values N. Although it may not appear to be immediately useful, it is worth remembering that one of the main high level language for scientific modelling is FORTRAN which processes arrays of numbers in matrices and vectors. Since each W_N is fixed, the components in the matrix W are therefore constant making the calculating easier to perform.

13.4 The DFT in DADiSP

DADiSP has a DFT command to allow the spectrum of a discrete time domain signal to be derived. It is very similar to the commands with which you are already familiar, but nonetheless we shall work through an example to ensure your understanding of the process. First generate a waveform with three frequencies and a sampling frequency of 1 kHz,

W1: gsin(300,.001,244)-gcos(300,.001,322)+gcos(300,.001,102)

Now apply the DFT to produce two spectral sets, one for the real and one imaginary. The result is shown in Figure 13.3,

W2: real(dft(w1))

W3: imag(dft(w1))

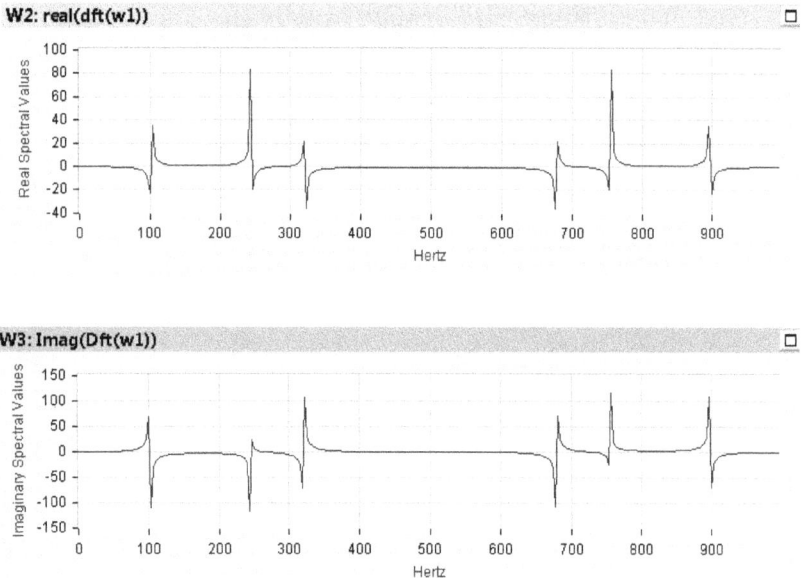

Figure 13.3: Real and Imaginary spectra from the DFT

You will observe from Figure 13.3 the spectral range is from 0 to 1kHz which is the sampling frequency. You will also observe the first half of the spectrum (0 → 500Hz) is an inverted mirror image of the second half of the spectrum (500Hz $\rightarrow$ 1kHz). The application of Eq:13.8 will form the magnitude spectrum. When the DFT is calculated in DADiSP it takes longer than the FFT which will be discussed in the next chapter. The DFT is however useful when you are interested in the behaviour of a few frequencies only. For example if you were monitoring 64Hz, 163 Hz and 240 Hz then with sample sets of 250 data points and a sampling frequency of 800Hz. The spectral bin width is $\frac{400}{125} = 3.2$Hz, therefore

$$k_{64} = \frac{64}{3.2} = 20, \ k_{163} = \frac{163}{3.2} = 51 \text{ and } k_{240} = \frac{240}{3.2} = 75$$

then,

$$X(64Hz) = \frac{1}{250} \sum_{n=0}^{n=249} x(n)e^{-j0.5026n}$$

$$X(163Hz) = \frac{1}{250} \sum_{n=0}^{n=249} x(n)e^{-j1.281n} \qquad ...13.19$$

$$X(240Hz) = \frac{1}{250} \sum_{n=0}^{n=249} x(n)e^{-j1.885n}$$

You can appreciate even for three frequencies there are many calculations to perform. There are systems where you need to measure the magnitude of a frequency which is proportional to another parameter you are measuring. An example can be found among some mechanical two axis gyroscopes where the rate of rotation in the two planes is proportional to *real[X(f₁)]* and *imag[X(f₁)]* where *f₁* is a fixed frequency. The DFT is therefore a useful process to measure this type of signal.

13.5 The Inverse Discrete Fourier Transform (IDFT)

Once a digital spectrum has been created, by performing the IDFT the result will be the digital signal in the time domain. The IDFT is useful after processing has been performed in the frequency domain and there is a need to return back the time domain data. An example is the MP3 file where data is stored in the frequency domain and an IDFT is used to reconstitute time domain audio track. The IDFT is expressed as,

$$x(n) = \sum_{k=0}^{k=N-1} X(k) \, e^{j\frac{2\pi nk}{N}} \qquad ...13.20$$

As an example of the application of the IDFT, simulate a spectral line in W1 of the form,

$$\frac{1}{(x-6)^2+0.001}$$

Click on f_x → **Generate Data** → **Y=F(X)**, when the diaglogue box opens enter the following,

Y=F(X): → 1/((x-6)^2+.001)
X Lower: → 0
X Upper: → 10
X Increment: → 0.01
Destination: W1.

When implemented in DADiSP the results are shown in Figure 13.4,

W1: 1/((x-6)^2+.001);sethunits('Hertz');setvunits('Magnitude')

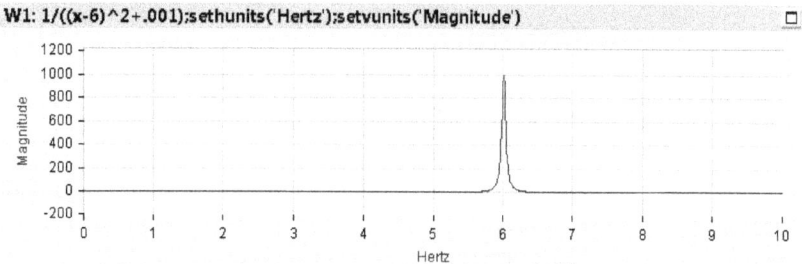

Figure 13.4: A spectral line at 6Hz

The IDFT of the spectral line shown in Figure 13.4 can be seen in Figure 13.5

W2: idft(w1)

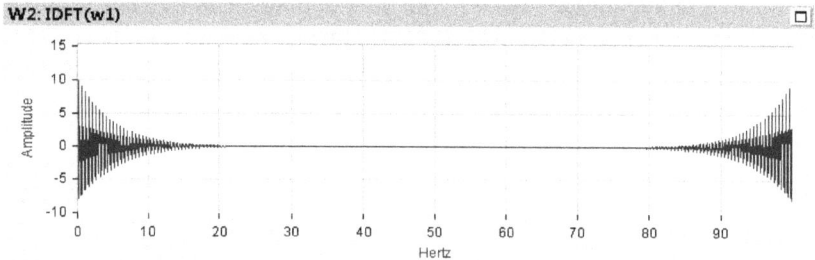

Figure 13.5: The result of performing a IDFT on Figure 13.4

The time domain trace in Figure 13.5 corresponds to a damped resonant waveform as expected from the IDFT. Incidentally, a real and imaginary data set are created. You will observe the mirror image of the waveform which starts after 50 seconds. In most examples the waveform after the first half is discarded. Although the DFT and its IDFT are effective tools for transferring between the time domain and the frequency domain, the computational load is quite heavy. In the next chapter the Fast Fourier Transform will be discussed which makes a significant saving in computational time.

13.6 The Discrete Power Spectrum

In Chapter 12 you were introduced to the concept of the *power spectral density* (PSD) and how it is related to the autocorrelation function. A digitised version of the PSD is given by,

$$P_{xx}(k) = \sum_{n=0}^{n=N-1} R_{xx}(n) \cos(\tfrac{2\pi n k}{N}) \qquad \text{...13.21}$$

where the discrete autocorrelation function (AFC) is given by

$$R_{xx}(n) = \tfrac{1}{M} \sum_{m=0}^{m=M-1} x(m)\, x(m+n) \qquad \text{...13.22}$$

There is however a problem when calculating the discrete AFC from finite length sequences of samples. Consider the example where the number of samples $M = 5$, expanding Eq:13.22,

$$R_{xx}(m) = \tfrac{1}{5}[x(m)x(m+1) + x(m)x(m+2) + ... + x(m)x(m+4)]$$

Expanding for all values of $0 \le m \le 5$,

$$R_{xx}(0) = \tfrac{1}{5}[x(0)x(0) + x(1)x(1) + x(2)x(2) + x(3)x(3) + x(4)x(4)]$$

$$R_{xx}(1) = \tfrac{1}{5}[x(0)x(1) + x(1)x(2) + x(2)x(3) + x(3)x(4)]$$

$$R_{xx}(2) = \tfrac{1}{5}[x(0)x(2) + x(1)x(3) + x(2)x(4)]$$

$$R_{xx}(4) = \tfrac{1}{5}[x(0)x(3) + x(1)x(4)]$$

$$R_{xx}(5) = \tfrac{1}{5}[x(0)x(4)]$$

You will observe as the lag value m increases, the number of data pairs making up the sum decreases since there are no values of $x(p)$ for $p > 4$. As each R_{xx} is divided by 5, the size of $R_{xx}(m)$ decreases with increasing m

value which brings an uneven weighting on $R_{xx}(m)$. There are various attempts to make the weighting even, for example,

$$\hat{R}_{xx}(n) = \frac{1}{M-n} \sum_{m=0}^{m=M-1} x(m)\, x(m+n) \qquad \dots 13.23$$

which can run into problems when M-n becomes very small. Using Eq:13.22, the new estimate of the autocorrelation function becomes.

$$\hat{R}_{xx}(n) = \frac{N}{M-n} R_{xx}(n) \qquad \dots 13.24$$

which is a closer estimate of the discrete AFC. So when performing the discrete PSD, you should be aware of potential problems relating to the calculation of the AFC.

What you have gained from this Chapter
1. Understanding the definition of the DFT and how it's derived from the Fourier Transform.
2. The concept of the frequency bin found in the digital frequency domain.
3. A method for calculating *sin(x)* and *cos(x)* from a truncated series.
4. How to calculate a square root of a number from an iterative process.
5. A method for calculating *tan⁻¹(x)* using a truncated series.
6. Representing the DFT in a matrix format.
7. How to implement the DFT in DADiSP.
8. The inverse DFT and its implementation in DADiSP.
9. The discrete AFC and the and the discrete PSD.

The DADiSP skill you have acquired from this Chapter
1. *gsin* and *gcos* for generating sin and cosine waveforms.
2. *real(DFT)* and *imag(DFT)* for obtaining the DFT of a waveform in a window.
3. $Y = F(X)$ - generating a general purpose function, in particular a spectral line.
4. *IDFT* - obtaining an inverse DFT.

14 Data Windowing

Although the Discrete Fourier Transform (DFT) produces a set of frequency bins, they are not perfectly rectangular, in fact each bin has the profile of a sinc function as illustrated in Figure 14.1. You will observe that each sinc profile bin spreads and overlaps into neighbouring bins. This is referred to as *spectral leakage*.

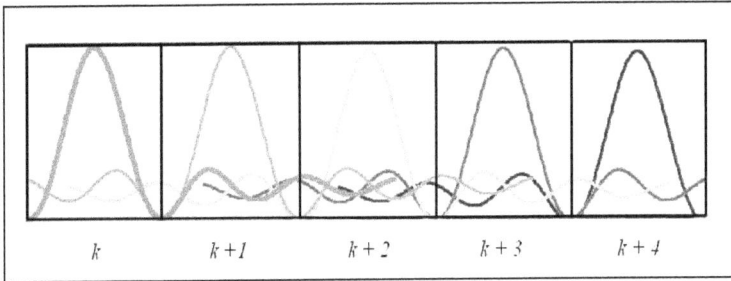

Figure 14.1: Actual profile of the frequency bins - not rectangular

You will recall the sinc profile has side lobes, it is these lobes which affect neighbouring frequency bins by leaking energy into them as shown in Figure 14.1. What is required is a method to minimise the leakage of energy from each spectral bin into neighbouring bins. This is where *data windowing* is used. The concept of windowing was introduced in Chapter 9 on FIR filter design. Applying a window is a straight forward process - it is a case of multiplication, if *x(n)* is the data batch and *w(n)* is the window function, the windowed data is,

$$x'(n) = x(n) \cdot w(n) \qquad \dots 14.1$$

If the original data set {*x*} had *N* data values, the data window also needs to have *N* data values which are either calculated in real-time or are stored and applied as required. For many windows the computational load is not too onerous, however others such as the Kaiser-Bessel, which will be covered later, the computational needed load is significant. We shall now look at the application of a data window. An example of the application of the *Hanning Window*, which has the form,

$$w_{Han}(n) = \tfrac{1}{2}[1 - \cos(\tfrac{2\pi n}{N-1})] \qquad \dots 14.2$$

W1: gsin(300,.01,10.2)

W2: ghanning(300,.01)

W3: w1*w2

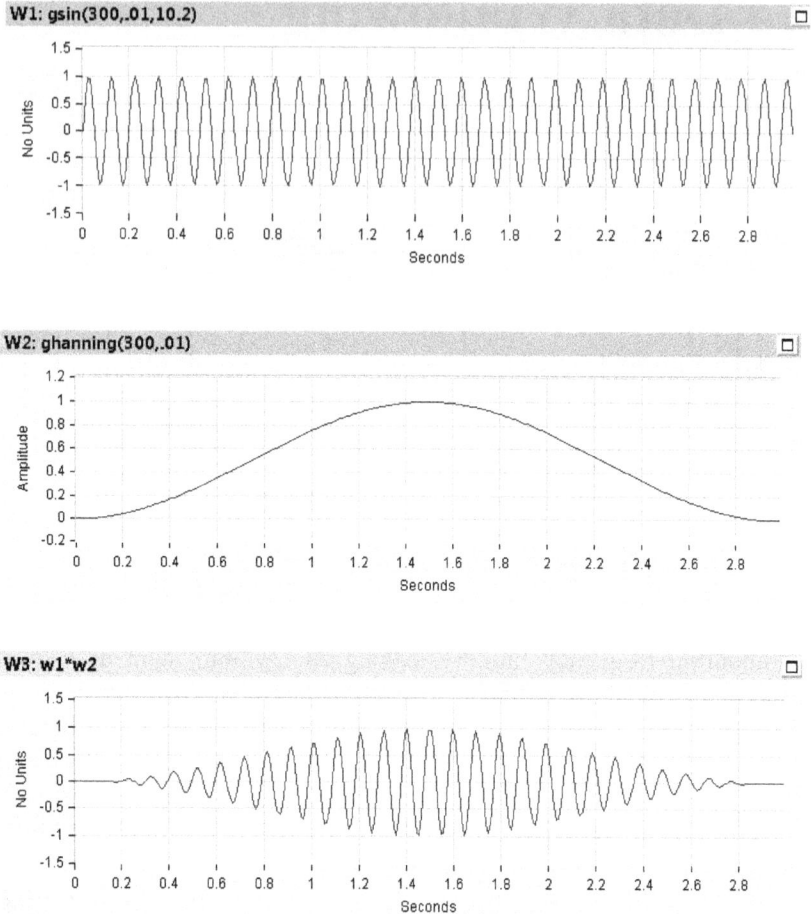

Figure 14.2: The effect of applying a window to a data batch

The effect of the Hanning Window is shown in Figure 14.2. You can see the data before windowing, the window function and the product. You will observe from Figure 14.2 the window discriminates against the data values at the start and end of the data batch *{x(n)}* - it reduces them to zero. The spectra of the data before and after the windowing is shown in Figure 14.3. You will notice from Figure 14.3 the spectrum of the data before windowing spreads across the whole spectral range - similar to a noise floor. Whereas the spectral peak of the windowed data is much sharper. The origin of the noise floor arises from the data points at the start and end of the data set.

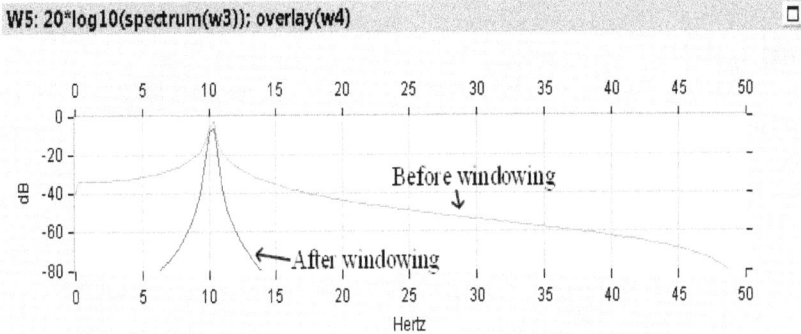

W5: 20*log10(spectrum(w3)); overlay(w4)

Figure 14.3: Spectral of data before and after windowing

If we were to take the signal in Figure 14.1 and wrap it round a cylinder so the start and end of the waveform connect we would get a waveform similar to that shown in Figure 14.4.

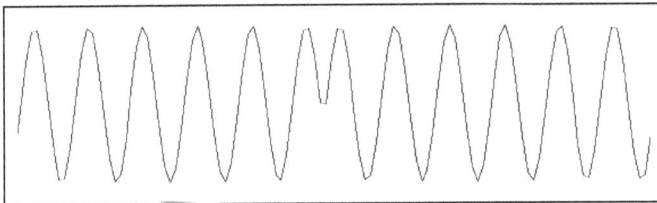

Figure 14.4: Mismatch at the start and end of data batch

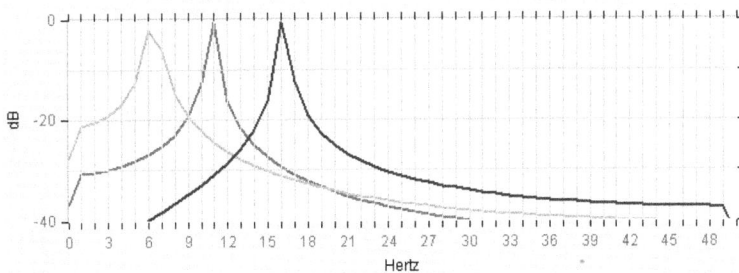

Figure 14.5: Overlay of three separate spectral

In Figure 14.4 you will observe a glitch or discontinuity in the sine wave. This is equivalent to injecting noise into the signal and this accounts for the noise floor which is evident in Figure 14.3. By applying the window the discontinuity is removed thereby removing the noise floor and giving a better spectrum. In Figure 14.5 you will see three separate spectra derived from three sine waves of different frequencies 6.4Hz, 10.8Hz and 16.2Hz. The spectra are not windowed and you will also observe the fifty spectral bins (the number of data points in each waveforms was 100). You can now appreciate how the energy from each spectral bin leaks into neighbouring spectral bins.

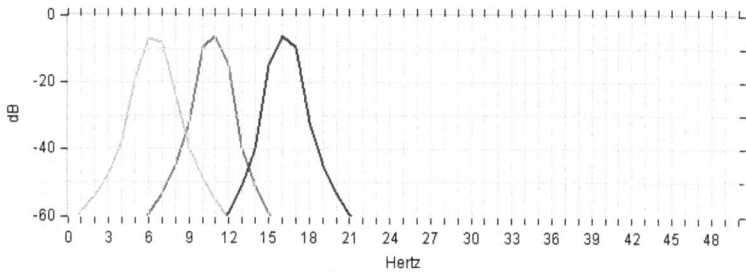

Figure 14.6: The effect of applying a window to each signal

Referring to Figure 14.6, once a Hanning window has been applied to each waveform you will observe a huge reduction in spectral leakage. Also note the dB scaling in Figure 14.6 which goes down to -60dB as opposed to -40 dB in Figure 14.5. At a magnitude of -20dB the leakage only occurs over three neighbouring frequency bins. The effect of applying a window is the broadening of the spectral lines and reasons for this effect have been given in Chapter 9.

Convolution in the Frequency Domain
In the discrete time domain when a window is applied to a data set {x} the resulting data set is according to Eq:14.1. In the previous section we have seen the effect of applying a window to data before the DFT has been applied. In the frequency domain, the equivalent process is,

$$X'(k) = \sum_{l=0}^{l=K/2} W(k) \, X(k-l) = W \star X \qquad \text{...14.3}$$

which is the spectrum of the window function W *convolving* with the spectrum X of the data. In Chapter 9 we considered the application of a window as modulation of half a cycle. The window function W in the frequency domain is a narrow band around 0Hz. When this is convolved with the spectrum of the data X, it is equivalent to W acting as a filter removing the low magnitude parts of the spectral profile. What remains is a refined spectral line - again make the comparison between Figure 13.7 and Figure 13.8. The several different window functions perform the same task with differing degrees of spectral refinement. You will also observe the secondary effect of applying the window causing the spectral line to broaden - the reason for this has already been given in Chapter 9.

14.1 Kaiser-Bessel Window

Among the commonly used windows the Kaiser-Bessel holds the distinction of having a variable β known as the *shape factor*.

$$W_{KB}(n) = \frac{I_0\left[\beta\sqrt{1-(\frac{n-a}{a})^2}\right]}{I_0(\beta)} \qquad ...14.4$$

where $a = (N-1)/2$ and I_o *is the zero order modified Bessel function of the first kind*. In fact I_o is given by the expression,

$$I_0(x) = 1 + \sum_{m=1}^{m=M}\left(\frac{(x/2)^m}{m!}\right)^2 \qquad ...14.5$$

In Eq:14.5 the value of M depends on the accuracy required to match the precision of the DSP on which the Kaiser-Bessel window is to be used. Performing real-time calculations is problematic, it is customary to save the window values in memory before hand. The significance of β is in shaping the window to the desired profile. Figure 14.7 shows the variation in the Kaiser-Bessel window as β varies between 2 and 12. By selecting the value of β any shape of window profile can be generated from the very gradual ($\beta = 2$) to the quite severe ($\beta = 12$). In practise in the Kaiser-Bessel window low values of $\beta = 2$ would not be used as their effect on the data values at the start and end of the sample batch would be very modest. The effect of applying the Kaiser-Bessel window to a data set is shown in the spectral in Figure 14.8. You will observe the different degrees of suppression on the periphery of the spectral line - in effect making the resolution more refined. Despite the increase in the value of β there is no significant increase in the spectral broadening.

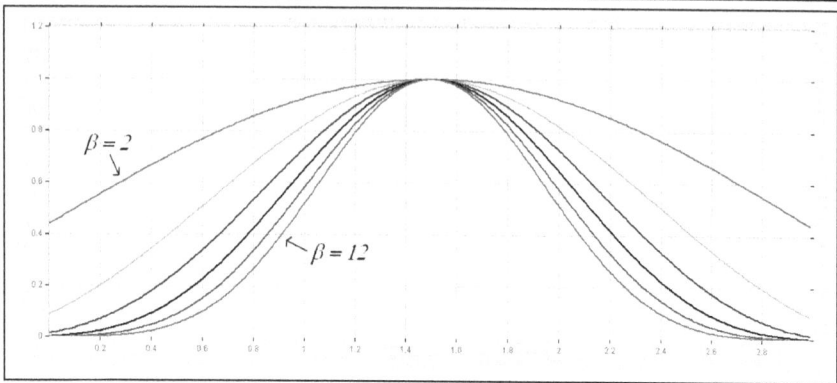

Figure 14.7: Kaiser-Bessel window profiles with the β = 2, 4, 6, 8, 10 and 12

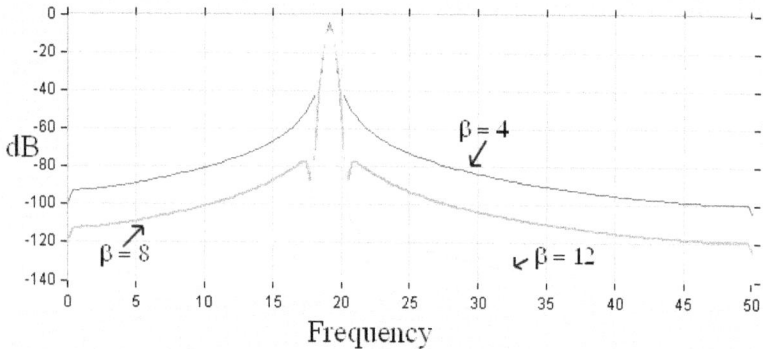

Figure 14.8: The effect of the Kaiser-Bessel window
with different values if β

Generally the Kaiser-Bessel Window lends itself more towards non-real-time processing owing to the heavy computational load needed for calculating each value $w_{KB}(n)$.

14.2 Flat-top Window
The majority of windows are good at allowing accurate measurements of frequency. The same cannot be said for measurements of magnitude. Consider the waveform shown in Figure 14.9.

W1: gsin(300,.01,6.5)+gsin(300,.01, 9.1)

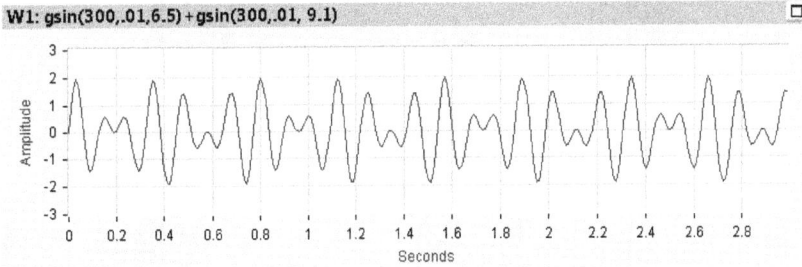

Figure 14.9: A waveform containing two frequency components

The spectrum of the waveform in Figure 14.9 is shown in Figure 14.10.

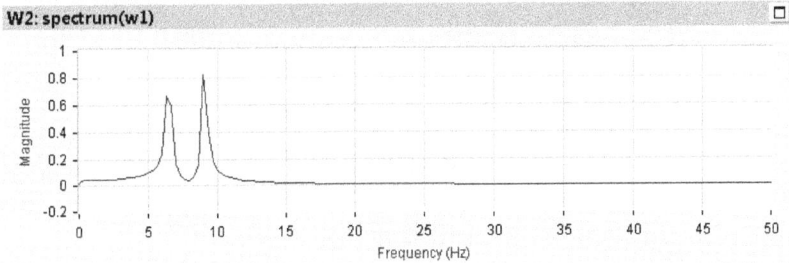

Figure 14.10: Spectrum of the waveform in Figure 13.12

You will observe in Figure 14.10 the two spectral lines, but they appear to have different magnitudes when in fact they have the same magnitude. This problem arises when spectral lines *fall between two stools*. Consider the two adjacent spectral bins - each having a sinc function profile. If two frequencies are processed and one frequency *f1* falls at the peak of one bin and the other *f2* at the crossover the bins there will be a difference in magnitudes as shown in Figure 14.11. You will observe from Figure 14.11 the two spectral lines occupying different positions in their respective frequency bins which gives *f1* the appearance of being larger than *f2* when in reality they have the same magnitude. This is referred to as the *Picket Fence* effect - when measuring the magnitude of spectral lines, the profile of the spectral bins can obscure their true magnitude. When using spectral analysis you need to be aware of its presence.

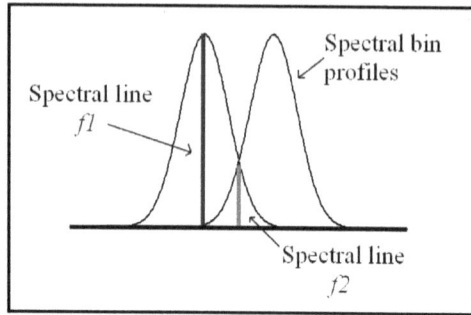

Figure 14.11: Two adjacent spectral bins with two spectral lines

To minimise this effect a *Flat-top* Window is used. It is expressed as,

$$w_{FT}(n) = 1 - 1.93\cos\left(\frac{2\pi n}{N-1}\right) + 1.29\cos\left(\frac{4\pi n}{N-1}\right)$$
$$- 0.388\cos\left(\frac{6\pi n}{N-1}\right) + 0.032\cos\left(\frac{8\pi n}{N-1}\right)$$

...14.6

The difference between the peak and the crossover point on two adjacent spectral bins is reduced to 0.01dB. The application of the flat-top window to waveform in Figure 14.9 is shown in Figure 14.12.

W1: (gsin(300,.01,6.5)+gsin(300,.01, 9.1))*gflattop(300,.01)

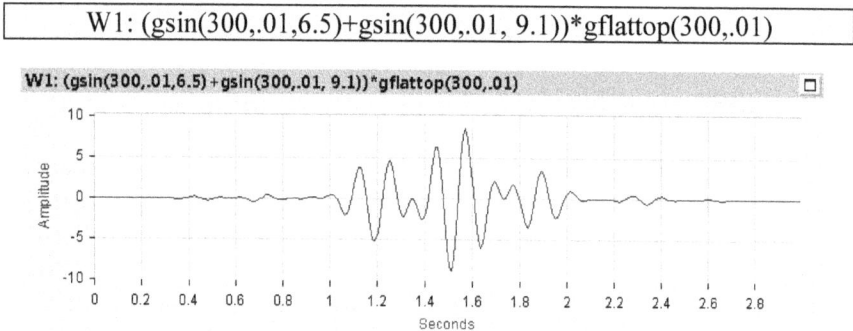

Figure 14.12: The application of the Flap-top Window

The spectrum of the windowed data is shown in Figure 14.13. You will observe from Figure 14.13 the heights of the two spectral lines are now the same as expected. You can verify this fact by evoking the **Vertical Cursor** (right mouse click) and moving the cursor across the peaks.

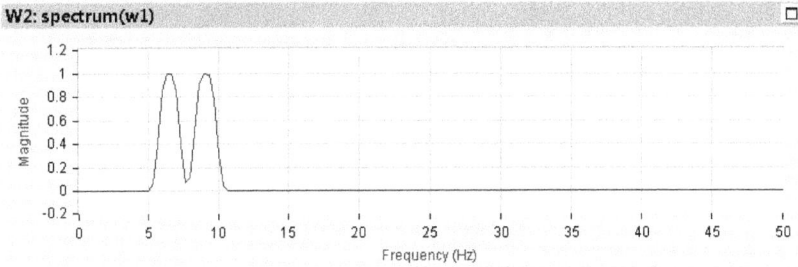

Figure 14.13: Spectrum after the application of a Flap-top window

You will also observe from Figure 14.13 a noticeable broadening of the spectral lines - a price you have to pay for gaining an accurate magnitude measurement. The majority of commercial digital spectrum analysers (see Figure 14.14) have this feature as there is a great need to measure the magnitude of spectral lines accurately.

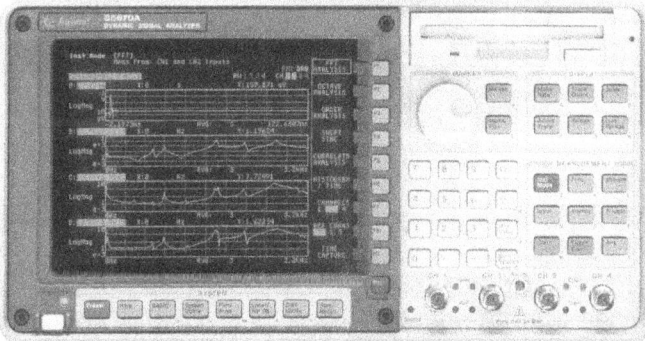

Figure 14.14: The Agilent 35670A Dynamic Signal Analyzer possesses a variety of Window functions including the Flat-top and Hanning Windows (Curtesy of Agilent Technologies).

14.3   Other Windows

Having established the need for data windowing before applying the Discrete Fourier Transform (DFT) it should be pointed out there are many data windows available from which to choose. When no data window is applied, this is equivalent to the application of a *rectangular data window* which is implicit in capturing a set of data samples from an analogue to digital converter - a sudden start and a sudden stop. Another well known window is the Hamming which is expressed as,

$$w_{Ham}(n) = 0.54 - 0.46\cos(\tfrac{2\pi n}{N-1})$$ ...14.7

In reality it is not too different from the Hanning Window. Both windows can be compared in DADiSP as shown in Figure 14.15.

W1: ghanning(300,.01)

W2: ghamming(300,.01)

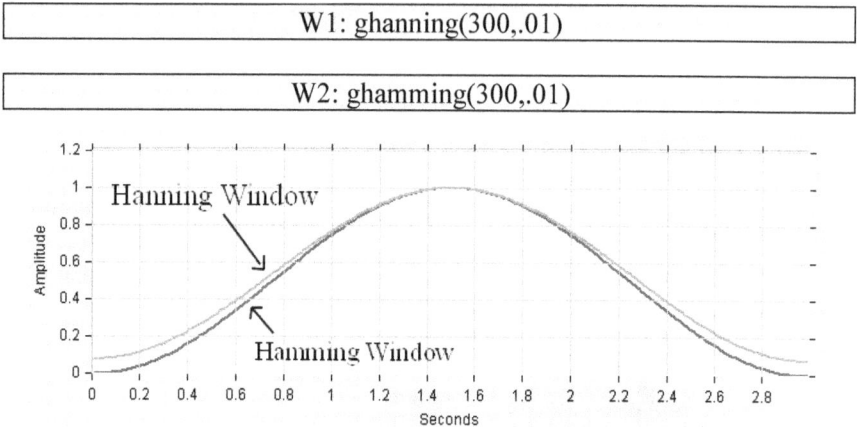

Figure 14.15: A comparison between the Hamming and Hanning Windows

You will observe from Figure 14.15 there is not at first sight a significant difference between the two Windows; although in reality the Hamming window has greater side lobe suppression. Another window which has been mentioned in the context of designing FIR filters is the Blackmann which is,

$$w_{Black}(n) = 0.42 + 0.5\cos(\tfrac{2\pi n}{N-1}) + 0.08\cos(\tfrac{4\pi}{N-1})$$ ...14.8

The Blackman window has an appearance not too dissimilar from other window which have been discussed as seen in Figure 14.16.

Table 14.1 shows a comparison between a number of commonly used windows. It shows the relative merits of each window and relates the broadening, that is the transition width and the side lobe suppression. In spectral analysis its customary to select a window depending on the degree of resolution required. Remember, the more you suppress the side lobes the greater the broadening effect of the window.

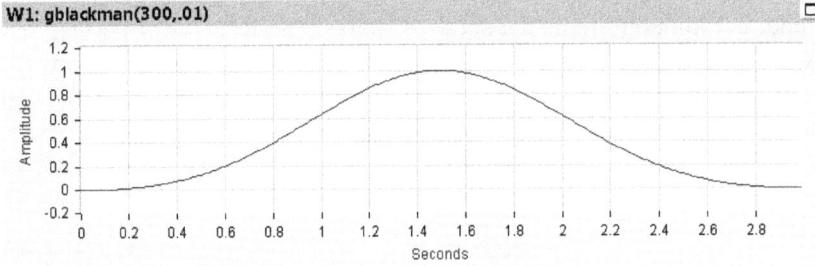

Figure 14.16: The profile of a Blackman window

Name	Transition Width	Maximum Side Lobe height dB
None (Rectangular)	0.9/N	21
Hanning	3.1/N	44
Hamming	3.3/N	53
Blackman	5.5/N	74
Kaiser-Bessel	2.93/N (β=4.52)	50
	4.34/N (β=6.76)	70
	5.71 (β=8.96)	90

Table 14.1: The relative effects of numerous windows

Example: It is believed there are two spectral lines separated by 80Hz and the smaller of the two is over 60dB smaller. What is the minimum number of data points needed to resolve the smaller spectral line if the sampling frequency is 20kHz ?

Solution: From Table 14.1, the window to choose is the Blackman owing to the side lobe height < 60dB. Therefore,

$$\frac{5.5}{N} = \frac{\delta f}{fs/2} = \frac{80}{10,000/2} = 0.016$$

Rearranging the expression,

$$N = \frac{5.5}{0.016} = 344 \text{ samples}$$

14.4 Near Perfect Spectral Resolution

Under certain conditions the spectral analysis performs near perfect resolutions. Referring to Figure 14.4, for most frequencies there is a poor match between the start and end of a batch of data samples. However for certain frequencies where there is a perfect match between start and end of the batch, the spectral resolution is greatly enhanced. If the samples are taken over a time T, the perfect match occurs for frequencies,

$$\frac{1}{T}, \frac{2}{T}, \frac{3}{T} \cdots \frac{n}{T} \qquad \qquad \dots 14.9$$

So is the value of T is 2 seconds, perfect resolution occurs at frequencies, 0.5Hz, 1Hz, 1.5Hz .... The highest frequency will depend on the sampling frequency. We can use DADiSP to simulate these perfect spectral resolutions. If we select a sine wave with 400 samples separated by 0.01 seconds - the duration of wave is 4 seconds. Now create a waveform with frequencies, 2Hz, 4Hz 8Hz and 16Hz which can be seen in Figure 14.17

W1: gsin(400,.01, 2)+gsin(400,.01, 4) +gsin(400,.01,8)+gsin(400,.01, 16)

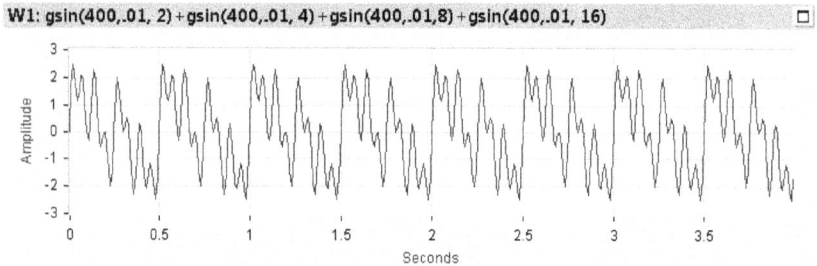

Figure 14.17: A waveform with an integral number of cycles

The waveform in Figure 14.17 has an integral number of cycles contained within the batch time T; perfect continuity therefore exists at the start and end of the sample batch. The spectrum of the waveform in Figure 14.17 is shown in Figure 14.18. In Figure 14.18 you will observe the four spectral lines present in the waveform in Figure 14.17. Notice they go beyond 250dB before the base begins to flatten out. This by any stretch of the imagination is very high spectral resolution, but unfortunately it only happens for relatively few frequencies.

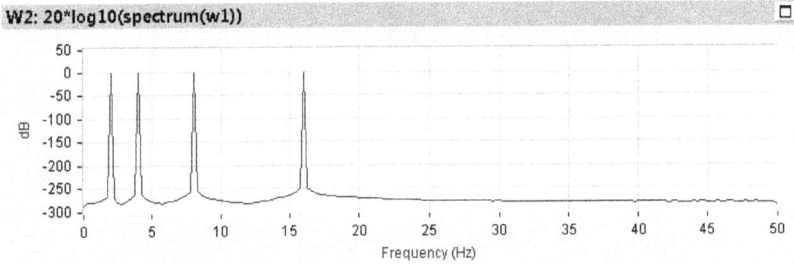

W2: 20*log10(spectrum(w1))

Figure 14.18: Spectral resolution with near perfect resolution

As an exercise change the value of one of the frequencies in W1 and observe the effect.

What you have gained from this chapter
1. How to window data in the time domain.
2. The effects of discontinuities at the start and end of a batch of data samples.
3. How spectral leakage affects the spectrum of a signal.
4. Beneficial effect of windowing is suppressing spectral leakage.
5. An understanding of convolution in the frequency domain.
6. Knowledge of the Kaiser-Bessel window.
7. Practise and knowledge of the Flat-top window.
8. The Hamming window in comparison with the Hanning window.
9. How to achieve near perfect spectral resolution.

The DADiSP skill you have acquired from this chapter
1. *gsin* - to generate a sine wave.
2. *ghanning* - to generate a Hanning Window.
3. *W1·W2* - to multiply two DADiSP windows together.
4. *gkaiser* - to generate a Keiser-Bessel Window with its shape factor.
5. *20*log10(spectrum))* - to obtain a dB spectrum.
6. *overlay* - to overlay two or more DADiSP windows over each other.
7. *gflattop* - to generate a Flat-top Window.
8. *ghamming* - to a generate a Hamming Window.
9. *gblackman* - to generate a Blackman Window.

15. The Faster Fourier Transform

Although the Discrete Fourier Transform in an effective method for obtaining the spectrum from a signal $x(t)$, it is not the most efficient. An algorithm for speeding up the calculation is the *Fast Fourier Transform* (FFT) which was realised by two American engineers J.W. Cooley and John Tukey in the 1960s, although it was first discovered by the famour mathematician *Carl Friedrich Gauss* around 1805. The advantage of the FFT over the DFT is the execution time - the FFT is much faster. In effect it arrives at the same answer by performing fewer calculations than the DFT. The DFT has a substantial number of redundant calculations, in fact it performs the same calculations more than once which is a waste of processing time. The FFT removes the repeated calculations which leads to a faster and more efficient algorithm. There is an added advantage of the FFT, when it is implemented on a Digital Signal Processor (DSP), since it performs the same task with fewer calculations, the accuracy of the spectral data it produces is greater - less rounding off of data values during the intermediate stages. This leads to a faster and more accurate algorithm. The discrete Fourier Transform is given by,

$$X_d(k) = \frac{1}{N} \sum_{n=0}^{n=N-1} x(n)\ W_N^{nk} \qquad \ldots 15.1$$

where

$$W_N = e^{-j2\pi/N} \qquad \ldots 15.2$$

which is often referred to as the *twiddle factor*. For illustrative purposes we shall consider an example of the DFT where the number of data values is 4 ($N = 4$) therefore,

$$X_d(k) = \frac{1}{4} \sum_{n=0}^{n=3} x(n)W_4^{nk}$$

$$= \frac{1}{4}[x(0)W_4^0 + x(1)W_4^k + x(2)W_4^{2k} + x(3)W_4^{3k}] \qquad \ldots 15.3$$

Now expand this expression for $k = 0, 1, 2, 3$,

$$X_d(0) = \frac{1}{4}[x(0)W_4^0 + x(1)W_4^0 + x(2)W_4^0 + x(3)W_4^0] \qquad \ldots 15.4a$$

$$X_d(1) = \tfrac{1}{4}[x(0)W_4^0 + x(1)W_4^1 + x(2)W_4^2 + x(3)W_4^3] \qquad \dots 15.4b$$

$$X_d(2) = \tfrac{1}{4}[x(0)W_4^0 + x(1)W_4^2 + x(2)W_4^4 + x(3)W_4^6] \qquad \dots 15.4c$$

$$X_d(3) = \tfrac{1}{4}[x(0)W_4^0 + x(1)W_4^3 + x(2)W_4^6 + x(3)W_4^8] \qquad \dots 15.4d$$

As you can see from Eq:15.4 for *4* data values there are *16* (= 4^2) complex multiplications and *16* (= 4^2) complex additions. This excludes the calculations for performing the sine and cosine operations. There are some 2×4^2 calculations. For an DFT with N samples there are in fact *of the order of $2N^2$ calulations*. This is expressed as $O(2N^2)$. When applying the FFT, the number of calculations reduces,

$$O(2N^2) \rightarrow O(N\log_2(N)) \qquad \dots 15.5$$

where N is the number of data values in the transform.

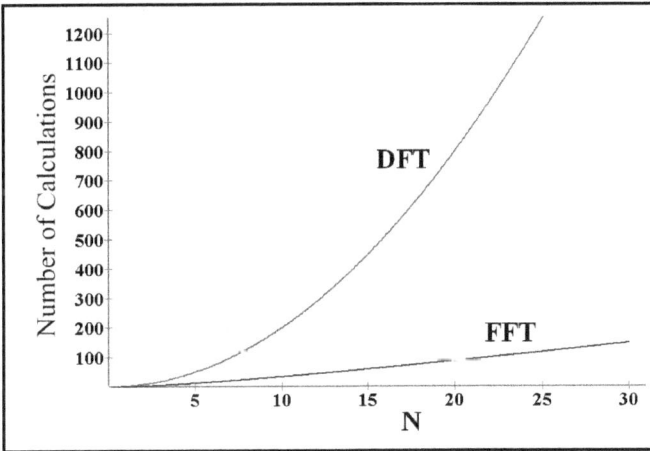

Figure 15.1: The difference between the number of calculations needed for the DFT and the FFT

Figure 15.1 shows the huge difference this makes in the number of calculations as N increases. You can appreciate that once the number of data values increases to 1000, the number of calculations reduces from 2,000,000 down to 9,966 - a substantial saving. The FFT is regarded as so important in the subject of digital signal processing that on some DSPs an FFT algorithm is embedded into the processors' silicon substrate. An early example is the ZR34161 from the *Zoran Corporation*.

The FFT Algorithm

The FFT algorithm partitions the data into odd and even data sets. If the data samples collected from an ADC is *{x(n)}* this is divided into two groups.

$$x_1(n) = x(2n) \text{ even samples}$$
$$x_2(n) = x(2n+1) \text{ odd samples}$$

...15.6

The Fourier Transform of these two sets are,

$$X_d(k) = \sum_{n=0}^{n=N/2-1} x(2n)W_N^{2nk}$$
$$+ \sum_{n=0}^{n=N/2-1} x(2n+1)W_N^{(2n+1)k}$$

...15.7

The division by N takes place at the end of the algorithm. The first term in Eq:15.5 can be written as

$$\sum_{n=0}^{n=N/2-1} x_1(n) \, W_{N/2}^{nk}$$

...15.8

And the second term in Eq:15.7 can be written as,

$$W_N^k \sum_{n=0}^{n=N/2-1} x_2(n) \, W_{N/2}^{nk}$$

...15.9

where $x_1 = x(2n)$ and $x_2 = x(2n+1)$. Therefore

$$X_d(k) = \sum_{n=0}^{n=N/2-1} x_1(n)W_{N/2}^{nk}$$
$$+ W_N^k \sum_{n=0}^{n=N/2-1} x_2(n)W_N^{nk}$$

...15.10

Now let,

$$X_{d1}(k) = \sum_{n=0}^{n=N/2-1} x_1(n)W_{N/2}^{nk}$$

and

$$X_{d2}(k) = \sum_{n=0}^{n=N/2-1} x_2(n)W_{N/2}^{nk}$$

...15.11

The Fourier Transform becomes

$$X_d(k) = [X_{d1}(k) + X_{d2}(k)W_N^k] \qquad ...15.12$$

for $k = 0, 1, ...N/2 - 1$. Now $X_{d1}(k)$ and $X_{d2}(k)$ are periodic, therefore for $k = N/2, N/2 + 1, ... N$

$$X_d'(k + \tfrac{N}{2}) = [X_{d1}(k) - X_{d1}(k)W_N^k] \qquad ...15.13$$

The minus sign in Eq:15.13 arises from,

$$W_N^{\left(k + \frac{N}{2}\right)} = e^{j2\pi\left(k + \frac{N}{2}\right)/N} = e^{j2\pi k/N} \, e^{j2\pi/2} = W_N^k \, (-1)$$

To calculate all the $X_{d1}(k)$ values you need to perform $2(N/2)^2$ complex operations. Likewise to calculate all the $X_{d2}(k)$ values you also need to perform $2(N/2)^2$ complex operations. Therefore to calculate $X_d(k)$ you need,

$$2(\tfrac{N}{2})^2 + 2(\tfrac{N}{2})^2 = N^2$$

The number of calculations has been reduced from $2N^2$ to N^2. You now partition the $x_1(n)$ samples into odd and even and also the $x_2(n)$ samples into odd and even and perform the same process. You keep on dividing down the data sets until you have *two data values* in each of the final sets. We shall consider the case where the number of samples is 4. Although this may appear a rather small data set, it is easier to appreciate the principles of the FFT with this small number. Therefore,

$$X_d(k) = x(0) + x(1)W_4^k + x(2)W_4^{2k} + x(3)W_4^{3k} \qquad ...15.14$$

When these are partitioned into even and odd groups we get,

$$X_{d1} = x(0) + x(2)W_4^{2k} \qquad ... 15.15$$

which becomes,

$$X_{d1} = x(0) + x(2)W_2^k \text{ for } k = 0 \qquad ... 15.16a$$

and

$$X'_{d1} = x(0) - x(2)W_2^k \text{ for } k = 1 \qquad ... 15.16a$$

Now let

$$X_{d2} = x(1) + x(3)W_2^k \text{ for } k = 0 \qquad ... 15.17a$$

and

$$X'_{d2} = x(1) - x(3)W_2^k \text{ for } k = 1 \qquad ... 15.17b$$

Therefore,

$$X_d(0) = [X_{d1} + X_{d2}W_2^0] \qquad \text{...15.18a}$$

and

$$X_d(2) = [X_{d1} + X_{d2}W_2^2] \qquad \text{...15.18b}$$

Also

$$X_d(1) = [X'_{d1} - X'_{d2}W_2^1] \qquad \text{...15.19a}$$

and

$$X_d(3) = [X'_{d1} - X'_{d2}W_2^3] \qquad \text{...15.19b}$$

Eq:15.16 to Eq:15.19 can be represented by using what are known as *butterflies*.

15.1 Butterfly Diagram

To construct a butterfly diagram we start with Eq:15.16 and Eq:17.

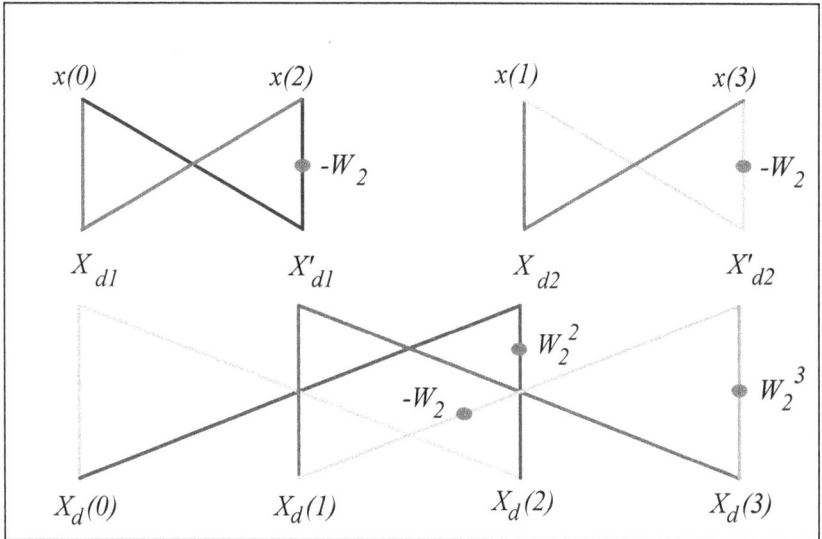

Figure 15.2: Butterflies for the an FFT when N = 4

You will observe from Figure 15.2 the number of stages required to arrive at the spectral set $X_d(k)$ is 2,

$$\log_2(N) = \log_2(4) = \log_2(2^2) = 2\log_2(2) = 2.$$

With a little imagination you can extend the model to greater number of data points. As mentioned the number of data points remaining in each decimated data set is 2. For this to occur it is necessary for the original data set to contain N samples where $N = 2^n$. Therefore for the FFT to work you need data sets which contain either $2^4 = 16$, $2^5 = 32$, $2^6 = 64$, $2^7 = 128$, $2^8 = 256$, ...Very often you will come across data sets containing $2^{10} = 1024$ values. Figure 15.2 also shows what is referred to as the *in-place algorithm* as you perform each stage of calculation, the results can occupy the same memory locations as the original data set. The final spectral data $\{X\}$ also occupies the same memory locations as the original data set $\{x\}$. In the 1960s memory was very expensive and every effort was made to use it efficiently. It should be remembered that each stage of the calculation requires twice as much memory - the need to hold the *real and imaginary* values in the spectral calculations. When the number of data points is decimated to 2 values per set this is referred to as the *radix-2 algorithm*. In Figure 15.2 you will observe the positioning of the twiddle factors and the role they play in deriving the final spectral set $\{X_d\}$.

Bit Reversal

When the FFT is performed in the manner as demonstrated in the previous section where the odd and even values are partitioned into different groups, this is referred to as *decimation in time*. Very often when the FFT is performed the raw data is just taken sequentially without separating into odd and even sample groups - this is referred to a *decimation in frequency*. When this happens the final spectral set $\{X\}$ is mixed up - the wrong spectral data values are in the wrong memory addresses. The method used to reposition the spectral data values in the correct order is called *Bit Reversal*. It operates on the principle shown in Figure 15.3. Representing the bit addresses in binary, each bit is replaced by its mirror image bit. Figure 15.3 shows 3-bit, 4-bit and 5-bit; bit reversal is applicable for any number. Once the bits have been reversed the new addresses are now the correct sequential locations for the spectral data values.

You will find that coded versions for the FFT in a high level language such as C++ will have a routine at the end of the program for performing the bit reversal. Whereas in DSP assembly langues, there is usually an instruction for performing this operation with relative ease.

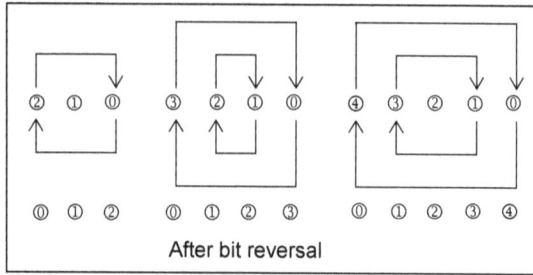

After bit reversal

Figure 15.3: Bit reversal, binary digits changing places

15.2 Zero Padding

Sometimes the requirement for the number of data points $N = 2^n$ cannot be realised, simply the sample batch does not match this number - it does not have enough samples. This can happen very easily as samples are often derived from an analogue to digital converter (ADC). When this happens a method known as *zero padding* is used to make up the number of data values to 2^n. This might sound rather surprising as it is not immediately obvious what the effects on the spectrum will be. For example if you collect 452 data values from an ADC, you are 60 data values short of 512. To make up the shortfall, 60 zero data values are added to the end of data set as shown in Figure 15.4.

W1: gsin(452,.001,87)

W2: gline(60,.001,0,0)

W3: concat(w1,w2)

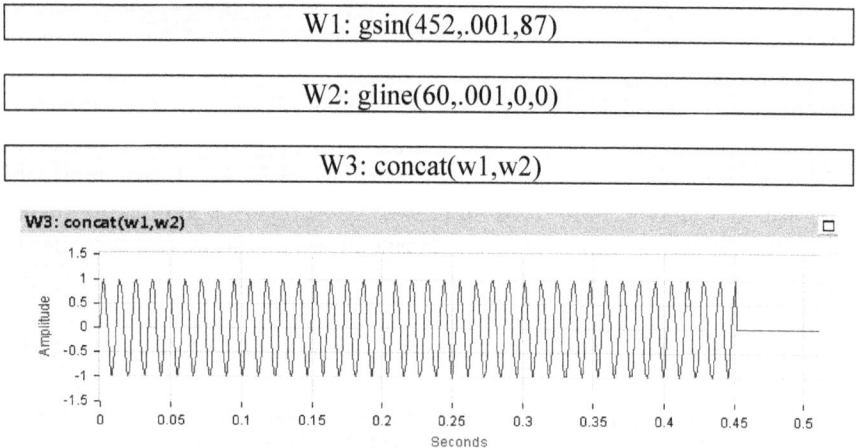

Figure 15.4: Adding on zeros to a waveform

Notice from Figure 15.4 the zeros have been added onto the end of the data values. The magnitude of the FFT spectrum of the waveform in Figure 15.4 is shown in Figure 15.5,

W4: 20*log10(real(FFT(w3))^2+Imag(FFT(w3))^2)

There are a number of interesting observations to be made from Figure 15.5; there is ripple on both sides of the spectral peak. Since 60 zero values are added to the waveform, the total number of frequency bins has been increased by 30.

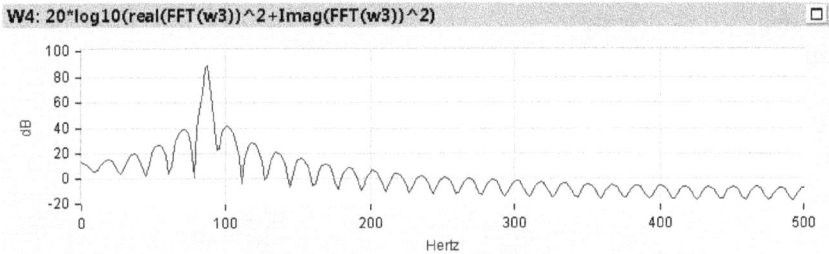

Figure 15.5: Spectrum of waveform in Figure 15.4

But there is no new frequency information to go in them. Hence the appearance of locations in the spectrum where there are bins with very little in them (they appear as zeros). There will however be some residual data in them resulting from spectral leakage. Once a Window is applied to the waveform in Figure 15.4, the ripple is removed as shown in Figure 15.6,

W3: concat(w1,w2)*ghanning(512,.01)

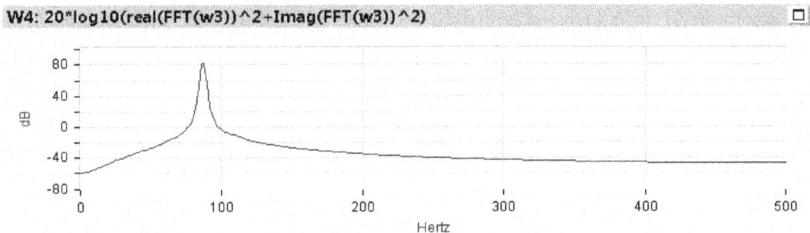

Figure 15.6: The effect of windowing the data before the FFT

207

You will observe from Figure 15.6 the zeros (the ripple) have been removed from the spectrum thus giving a better defined spectral line - compare the magnitude of Figure 15.5 with Figure 15.6, the latter has far better definition. When you confirm the trace in Figures 15.5 & 15.6 you will discover a *mirror image* of the spectrum starting at 500 Hz. This is a consequence of performing an FFT according to Eq:15.13 - the peridicity of $X_{d1}(k)$ and $X_{d2}(k)$. It is customary to apply data windows before the application of the FFT, in a similar manner to that of the Discrete Fourier Transform. The information relating to data windows in the previous chapter is equally relevant when applying the FFT. The application of data windows only becomes questionable when a signal has a high noise content, in which case little is gained from windowing apart from the undesirable effect of broadening the spectral lines.

15.3 The FFT and Complex Input Data

Normally the data on which the FFT performs is real, however it is equally capable of performing on complex data. Consider the Fourier Transform,

$$X_{xy}(k) = \frac{1}{N} \sum_{n=0}^{n=N-1} [x(n) + jy(n)] \, W_N^{nk} \qquad \ldots 15.20$$

You will observe from this expression the input data has two data sets *{x(n)}* and *{y(n)}*.

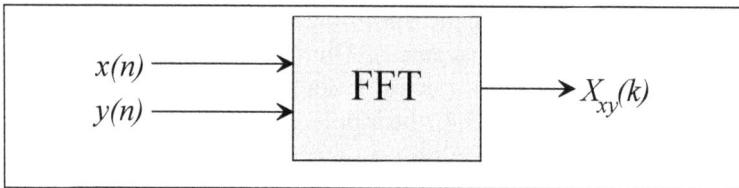

Figure 15.7: The FFT processing two input data streams

As illustrated in Figure 15.7 the FFT can process two input data streams and an obvious question is how this affects the spectrum $X_{xy}(k)$. You will remember when you repeated the trace in Figures 15.5 & 15.6 how mirror images appeared in the spectrum. The FFT from Eq:15.20 does not produce mirror images which will now be demonstrated. First create two waveforms containing different frequencies,

W1: gsin(512,.001,314)-.5*gcos(512,.001,115)

and

> W2: gsin(512,.001,56)-0.4*gsin(512,.001,22)

There are frequencies are at 314Hz, 115Hz, 56Hz and 22Hz. These are combined to form a complex waveform,

> W3: complex(w1,w2)

The magnitude of FFT spectrum is generated by,

> W4: 20*log10(real(FFT(w3))^2+Imag(FFT(w3))^2)

which appears as shown in Figure 15.8,

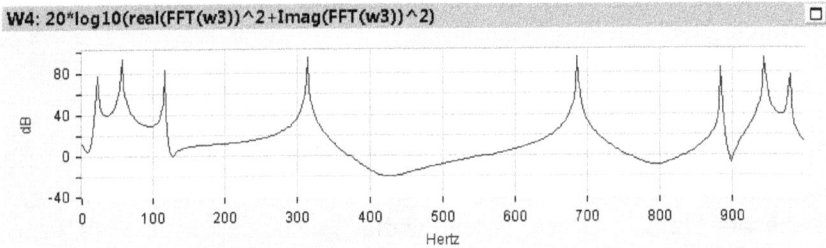

Figure 15.8: Spectrum of complex waveform in W3

You will notice the spectrum no longer possesses a mirror image symmetry as before. The sampling frequency is 1,000 Hz and what you see beyond 500 Hz rightly belongs in the negative part of the spectrum. In effect the spectrum beyond 500Hz should be *translated* to the left by 1,000Hz. The attractive feature of this processing is the ability to observe the spectra of two separate waveforms using one FFT only.

What you have gained from this Chapter

1. The FFT and how the data is partitioned into separate odd and even groups.

2. Decreasing the number of calculations from $2N^2$ to $N\log_2(N)$.

3. The role played by the Butterfly diagrams in the understanding of the FFT.

4. The need for bit reversal in FFTs.
5. The need for zero padding to make the FFT possible.
6. The effect of zero padding on spectra.
7. Using the FFT on two data sets simultaneously.
8. Affect on the spectrum when using complex input data.

The DADiSP skills you have acquired from this Chapter

1. *gsin* - to generate a sine wave.
2. *gline* to generate a straight line, with or without a gradient.
3. *concat* - to concatenate two or more waveform - joining them end to end.
4. *FFT* - to obtain a spectrum from a waveform
5. *ghanning* - to apply a Hanning data window
6. *complex* - to convert two waveforms into a single complex waveform having real and imaginary data sets.

DADiSP Extra

One of the special functions in DADiSP is the *sweep* where the waveform starts at one frequency and ends with a higher frequency. Click on $f_x \rightarrow$ Generate Data $\rightarrow$ Other. When the dialogue box opens; Series Type $\rightarrow$ SWEEP, Num. Points: $\rightarrow$ 1024, Spacing: $\rightarrow$.01, Start Freq: $\rightarrow$ 2, End Freq: $\rightarrow$ 10 Destination: W3. The sweep waveform its spectrum is shown in Figure 15.9.

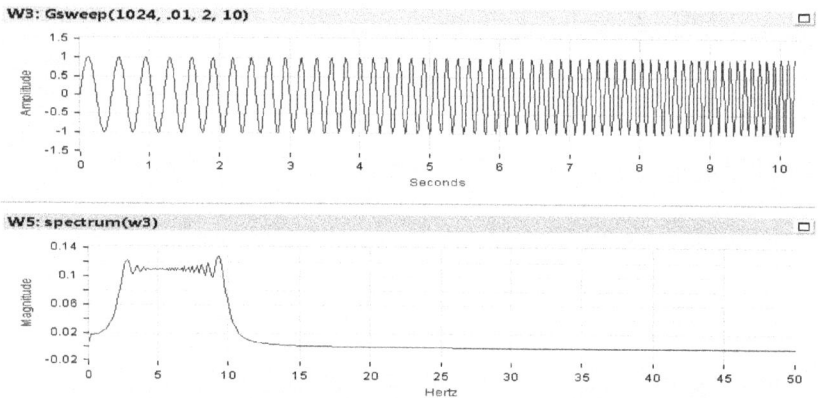

Figure 15.9: A sweep waveform and its spectrum

16. Final Thoughts

If you have managed to progress through all the chapters and have confirmed the DADiSP simulations, then congratulations are in order - well done. You are now in a position to make further strides into the highly rewarding subject of digital signal processing. By now you will have realised that DADiSP is a very powerful package and it would be in your interest to buy a full version of the software which is available from,

www.dadisp.com

Having progressed through this book you would have arrived at the inescapable conclusion that digital signal processing requires a high level of mathematical knowledge and skill. And as further research is carried out in the subject, the level of mathematics will only increase. Therefore to be a proficient engineer it is necessary for you to maintain and if possible improve your mathematical skills. As explained right at the beginning of the book, you should now be in a strong position to make full use of the many excellent text books which have been published on this subject. The knowledge you now posses will enable you to gain further understanding and advancement in digital signal processing from many of these expertly written text books (see the end of this chapter).

16.1 Future topics

The material covered in this book is rather modest compared with the wealth of knowledge which exists in the domain of digital signal processing. As you make further progress in the subject, you will invariably come across some of the following;

- Lattice filters
- Decimation Filters
- Mirror Filters
- Adaptive filters and adaptive algorithms - both non-recursive and recursive
- The Hilbert Transformer
- Multirate processing
- Kalman filters for performing in real-time with high levels of noise and signal drop-out

- Modern spectral analysis using parametric spectral algorithms
- Radix-4 and Radix-8 FFT algorithms
- Winograd algorithms for spectral analysis
- Cepstrum Analysis
- Wavelet Analysis
- Data compression
- Error correction and encryption
- Digital signal Processor architecture

As you can appreciate from this list there is still lots more to learn about in the field of digital signal processing, but at least you now have the fundamental tools and understanding to make good progress in these other areas of DSP knowledge. In the several books referenced in Chapter 1 and further in this chapter you will find all of the above list of subjects covered in some depth.

16.2 Future Development

The other side of digital signal processing is to implement algorithms in real-time. This usually involves using digital signal processors such as a member from the *Texas Instruments TMS320 family*. Alternatively you may even have the opportunity to implement a DSP algorithm on a *Field Programmable Gate Array* (FPGA) IC such as the the *Xilinx Virtex-6*. Whichever solution you pursue, there will be a new learning curve for you to follow; assembly language in the former or VHDL in the latter. Nonetheless acquiring these new skills will add to your employ-ability which is good news for your future. Although a full discussion of the DSP development tools is outside the scope of this book, a brief insight will allow you ask the right questions in the future when you begin to focus your attention on real-time appli-cations. After all digital signal processing is only really useful when it is implemented in real-time systems. The hardware and software tools you will encounter when you enter the world of DSP development include,

- *Assembler* - a software utility for converting assemble source code (your program) into a hexadecimal code module. Every DSP has an assembly language which is used to implement (code) algorithms on the processor.

- *Linker* - a software utility for linking several coded modules together, assigning memory address values and producing an executable binary code - the program to run on the DSP.

- *Library* - a collection of frequently used modules which can be linked into your code.

- *Evaluation Module* (EVM) - a hardware PCB hosting a DSP IC which contains it own memory containing a software *monitor*. This allows direct communication between a PC and the EVM and includes features for downloading executable code from the PC to the EVM. You can then observe the execution of the code line by line as the monitor sends back to the PC the contents of the processor's registers. Several EVMs have peripheral ICs such are ADCs and DACs to enable real-time program execution. An example of such an EVM is shown in Figure 16.1 which hosts the Motorola DSP56307 processor.

Figure 16.1: The Motorola DSP56307 EVM

- *Incircuit emulator* (ICE) - a hardware device which has a pod that replaces and replicates the DSP on a prototype circuit board during the testing and debugging phase of development. Once a DSP IC has been inserted into the prototype board there is no means of testing the true functionality of the software, hence the ICE is therefore used to test

213

and debug the software before the actual processor is inserted into the prototype board. The ICE is usually linked up to a PC which is able to monitor the activity in the processor.

- *Logic Analyser* - an instrument with multiple input channels that allows the monitoring of bus activity on a microprocessor or DSP. It would normally be used for fault finding after most of the programs and hardware have been tested and debugged. Often used to identify intermittent glitches which occur when there are timing issues on a circuit board.

This is a very brief overview of a few development tools which are used in the design and development of DSP products. It allows you to be aware that such tools exist and in the future you will probably become acquainted with several of them as your education and career advances into the field of digital signal processing.

16.2 Using Real-time Spectrum Analysers

It is quite possible you will be expected to put your knowledge of signal processing into practise by performing direct measurements and interpreting the results. To achieve this objective you will probably use a commercial real-time FFT analyser such as the *Agilent 35670A* as shown in Figure 14.14. This instrument has a huge wealth of functions which will be detailed in its User's Manual. Before you attempt to use one of these instruments it is useful to have an insight into the range of operations they perform. Typically their spectral range goes up to 100kHz and they will have multiple input channels. Table 16.1 shows a very brief list of some of the signal processing functions which can be performed by the 35670A.

FFT analysis	Histogram/time	Correlation analysis
Time capture	Frequency response	Power spectrum
Linear spectrum	Coherence	Power spectral density
Phase Polar plot	Autocorrelation	Cross-correlation Orbit diagram
Histogram, PDF	Linear magnitude	Time waveform Amplitude domain
Group delay	dB magnitude	Log magnitude
Imaginary part	Real part	phase

Table 16.1: Some of the functions on the Agilent 35670A spectrum analyser

Having read through this book you will recognise the functions listed in Table 16.1 - the type of operations with which a practising electronic and mechanical engineer should be familiar. As you can see from Figure 14.14 the 35670A has four input channels (four BNC connectors at the base) and amongst the many useful functions which are available is the *waterfall display* as shown in Figure 16.2.

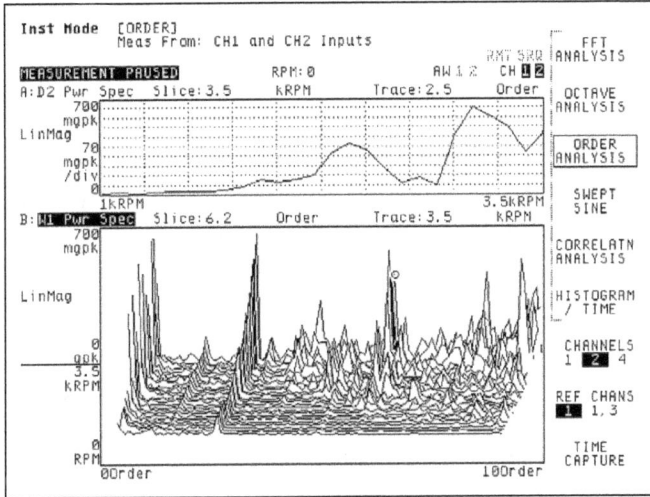

Figure 16.2: A waterfall display

The waterfall feature is made up of partially overlapping spectra, as each new spectrum is calculated, the previous spectrum is displaced slightly to the right and pushed back which in turn pushes back the previous spectra. This creates the impression of a waterfall process in reverse. This feature is very useful for monitoring the changing spectral content of a signal. Its direct application would be found when performing dynamic vibration measurements on rotating machines. You will also find on the right hand side of the Figure 16.2 various functions which are activated by the *soft keys* on the instrument (see Figure 14.14). Typically real-time analysers also provide various output signals which serve as inputs for system testing. These include *random burst, random, periodic chirp, burst chirp, pink noise* and *sine*. For mechanical system testing these signals will need amplification.

This is a very brief glimpse at the range of features found on a commercial real-time spectrum analyser, many of the features you will already have encountered in the DADiSP simulations in this book.

Further reading

Here is a list of books on signal processing which are available on-line or from your university library or even from a local library through inter-library loan scheme. All these books are worth visiting as you will find the information in each valuable and should contribute further to your under-standing of digital signal processing.

Introduction to Digital Signal Processing by Johnny Johnson, Prentice-Hall International Edition (ISBN: 0134806344).

Discrete Random Signals and Statistical Signal Processing by Charles Therrien, Prentice-Hall (ISBN: 0138521127).

Digital Signal Processing: Spectral Computation and Filter Design by Chi-Tsong Chen, Oxford University Press (ISBN: 0195136381).

Introduction to Discrete-Time Signals and Systems by R.I. Damper, Chapman & Hall (ISBN: 0412476501).

Introduction to Digital Signal Processing by Roman Kuc, McGraw-Hill (ISBN: 0071005439).

Digital Signal Processing: Concepts and Applications by B. Mulgrew, P. Grant and J. Thompson (Second Edition) Palgrave MacMillan (ISBN: 0333963463)

Digital Signal Analysis, by S Stearns and D Hush, Prentice-Hall (ISBN: 013211772X)

Signal Processing and Linear Systems by B.P. Lathi, Berkeley Cambridge Press (ISBN: 0941413357).

Understanding Signals and Systems by Jack Golten, McGraw-Hill (ISBN: 0077093208).

Linear Systems and Signals, B.P. Lathi, Oxford University Press (ISBN: 0195151291).

A Digital Signal Processing Primer: with Application to Digital Audio and Computer Music by Ken Steiglitz, Addison-Wesley (ISBN: 0805316841).

C Language Algorithms for Digital Signal Processing by P. Embree and B. Kimble, Prentice-Hall (ISBN: 013133406-9).

Discrete-Time Speech Signal Processing: Principles and Practice by Tomas Quatieri, Prentice Hall (ISBN: 013242942X).

Introduction to Digital Signal Processing by J. Proakis and D Manolakis, Macmillian (ISBN: 002396815X).

Singnal Processing: A Modern Approach by J. Candy McGraw-Hill (ISBN:0071004106).

Digital Signal Processing by D.DeFatta, J. Lucas and W. Hodgkiss, John Wiley (ISBN: 0471637653).

An Introduction of Random Vibrations, Spectral & Wavelet Analysis by D.E. Newland (Third edition) Longman (ISBN: 0582215849).

Random signal Processing by Dwight Mix, Prentice Hall (ISB:N0131801912).

Introduction to Random Signals and Applied Kalman Filtering by Robert Grover Brown and Patrick Hwang, John Wiley (ISBN: 9780471559221).

Statistical Digital Processing and Medeling by Monson Hayes John Wiley & Sons (ISBN: 0471594314).

Introduction to Statistical Signal Processing with Applications by M.D. Srinath, R.K. Rajasekaran and R. Viswanathan, Prentice Hall (ISBN: 0131252950).

Fast Fourier Transform and Convolution Algorithms by H.J. Nussbaumer, Springer-Verlag (ISBN: 038710159).

Digital Signal Processing Laboratory by B.Preetham Kumar, CRC Press (ISBN: 9781439817377).

Digital Signal Processing: DSP & Applications by Dag Stranneby, Newnes (ISBN: 0750648110).

Digital Signal Processing: A Practical Guide for Engineers and Scientists by Steven Smith, IDC Technology, (ISBN: 0750674447).

Appendix: Partial Fractions and DADiSP

Partial Fractions are used in the understanding of Infinite Impulse Response (IIR) filters. Given two polynomials $P(x)$ and $Q(x)$, the ratio of these may be expressed as a sum of fractions,

$$\frac{P(x)}{Q(x)} = \frac{c_1}{x-a_1} + \frac{c_2}{x-a_2} + \ldots + \frac{c_n}{x-a_n} = \sum_{m=1}^{m=n} \frac{c_m}{x-a_m} \qquad \ldots \text{A1}$$

The values of $\{c\}$ can be determined by equating coefficients in terms of the power of x. The expansion in Eq:A1 is only possible if the order of $P(x)$ is greater than the order of $Q(x)$ otherwise the ratio can be expressed as,

$$\frac{P(x)}{Q(x)} = D(x) + \frac{R(x)}{Q(x)} \qquad \ldots \text{A2}$$

Partial fractions often fall into one of the three following categories.

1. *Linear factors*

$$\frac{C}{(x-a_1)(x-a_2)} = \frac{c_1}{x-a_1} + \frac{c_2}{x-a_2} \qquad \ldots \text{A3}$$

2. *Containing a quadratic factor*

$$\frac{Cx}{(x-a_1)(x^2+a_2)} = \frac{c_1}{x-a_1} + \frac{c_2x+c_3}{x^2+a_2} \qquad \ldots \text{A4}$$

3. *Has a repeated factor*

$$\frac{Cx}{(x-a_1)(x+a_2)^2} = \frac{c_1}{x-a_1} + \frac{c_2}{x+a_2} + \frac{c_3}{(x+a_2)^2} \qquad \ldots \text{A5}$$

Trying to derive the various coefficients is rather tedious and fortunately DADiSP has two commands for performing this task depending whether the partial fractions are required in S plane (*residue*) or the Z plane (*residuez*). We shall work through a few examples, firstly using the *residue* command.

Example 1:   Consider the expression,

$$H(s) = \frac{1}{s^2+4s+3} \qquad \ldots \text{A6}$$

Enter the coefficients into the *residue* command,

W1: residue({1}, {1, 4, 3})

The results are as shown in Table A.1 (you may have to use the F7 key several times to reach the table display),

W1: residue({1}, {1, 4, 3})

	1: Residues	2: Poles
1:	-0.500000	-3.000000
2:	0.500000	-1.000000
3:		

Table A.1: Results from residue *using Eq: A6*

The results from Table A.1 are used to complete the expansion of Eq:A6,

$$H(s) = -\frac{0.5}{(s+3)} + \frac{0.5}{(s+1)} \qquad \dots \text{A7}$$

Example 2: Consider the expression,

$$H(s) = \frac{s^2+s+1}{s^3-5s^2+8s-4} \qquad \dots \text{A8}$$

Enter the coefficients into the *residue* command,

W1: residue({1, 1, 1}, {1, -5, 8, -4})

The results are as shown in Table A.2,

W1: residue({1, 1, 1}, {1, -5, 8, -4})

	1: Residues	2: Poles
1:	3.000000	1.000000
2:	-2.000000	2.000000
3:	7.000000	2.000000
4:		

Table A.2: Results from residue *using Eq: A8*

The results from Table A.2 are used to complete the expansion of Eq:A8,

$$H(s) = \frac{3}{(s-1)} - \frac{2}{(s-2)} + \frac{7}{(s-2)^2} \qquad \dots \text{A9}$$

Note what happens when repeated root occurs (refer to Eq:A5). Having looked at the residue we shall now look at *residuez* in the Z domain.

Example 3: Consider the expression,

$$H(z) = \frac{2+3z^{-1}+4z^{-2}}{1+3z^{-1}+3z^{-2}+z^{-3}}$$ ... A10

Enter the coefficients into the *residuez* command,

W1: residuez({2, 3, 4}, {1, 3, 3, 1})

The result are shown in Table A.3,

W1: residuez({2, 3, 4}, {1, 3, 3, 1})							□
	1: Residues	2: Poles					
1:	4.000000	-1.000000					
2:	5.000000	-1.000000					
3:	3.000000	-1.000000					
4:							

Table A.3: Results from residuez using Eq: A10

From this table we see again what are known as the residue values and the three repeated poles at $z = 1$. Eq:A6 is therefore expressed as,

$$H(z) = \frac{4}{1+z^{-1}} + \frac{5}{(1+z^{-2})^2} + \frac{3}{(1+z^{-1})^3}$$ ... A11

Note that since all the poles are located at $z = 1$, the three partial fractions are expressed as powers 1 to 3.

Example 4: Consider the expression

$$H(z) = \frac{0.1948z^{-1}}{1-1.3483z^{-1}-0.5986z^{-2}}$$ ... A12

Enter the coefficients in *residuez* command,

W2: residuez({0, 0.1948}, {1, -1.3483, -0.5986})

The results are shown in Table A.4.

W1: residuez({0, 0.1948}, {1, -1.3483, -0.5986})							□
	1: Residues	2: Poles					
1:	0.094914	1.700346					
2:	-0.094914	-0.352046					
3:							

Table A.4: Results from residuez using Eq:A12

From Table A.4, the expression in Eq:A12 becomes

$$H(z) = \frac{0.094914}{(1-1.700346z^{-1})} - \frac{0.0949145}{(1+0.352046z^{-1})} \qquad \dots A13$$

Example 5: Consider the expression,

$$H(z) = \frac{1+0.5z^{-1}+0.5z^{-2}}{1-0.1801z^{-1}+0.3419z^{-2}-0.0165z^{-3}} \qquad \dots A14$$

Enter the coefficients into the *residuez* command,

W3: residuez({1, 0.5, 0.5}, {1, -0.1801, 0.3419, -0.0165})

The results are shown in Table A.5.

W1: residuez({1, 0.5, 0.5}, {1, -0.1801, 0.3419, -0.0165})				□
	1: Residues		2: Poles	
1:	-0.295032	-0.556506i	0.065457	0.575479i
2:	-0.295032	0.556506i	0.065457	-0.575479i
3:	1.590064	0.000000i	0.049186	0.000000i
4:				

Table A.5: Results from residuez *using Eq: A14*

From Table A.5, the expression in Eq:A14 becomes

$$H(z) = \frac{(z+0.295031+j0.556506)}{(z-0.065457-j0.57549)}$$

$$+ \frac{(z+0.295032-j0.556506)}{(z-0.065457+j0.575479)} \qquad \dots A15$$

$$+ \frac{(z-1.590064)}{(z-0.04919186)}$$

The locations of the poles and zeros for Eq:A14 are shown on Figure A.1 from using the DADiSP *zplane* command. When considering Eq:A15, reference is no longer made to zeros, only poles. From Eq:A15 you will notice that *H(z)* → ∞ if either of the three denominators approach zero. Whereas *H(z)* → 0 does *not* occur if either of the nominator values approach zero - they are therefore not *zeros*. From these examples you should have gained a good understanding of how the commands *residue* and *residuez* are used to factorise expressions in both the S plane and the Z domain.

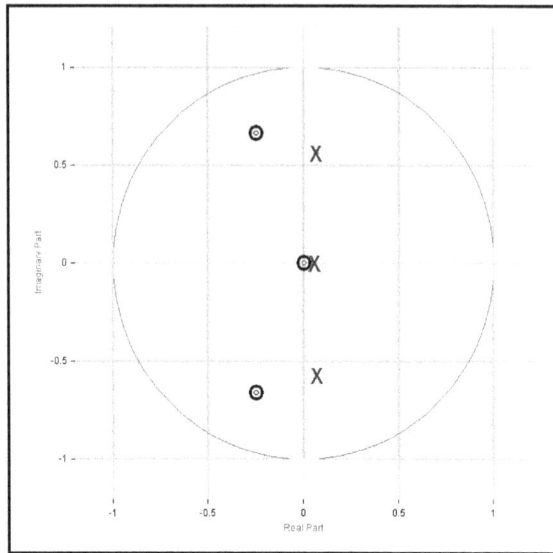

Figure A.1: The unit circle showing the poles and zeros in Eq:A14

Referring back to Eq:A2, if this expression is true, you can still use the *residue* command without having to perform any manual division. To illustrate this process we shall work through an example.

Example 6: Consider the expression,

$$\frac{P(x)}{Q(x)} = \frac{x^3-2x^2-3x+4}{x^2+4x-5} \qquad \dots A17$$

You will observe from this expression the order to the nominator is 3 whereas the order of the denominator is 2. Using the *residue* command the results are shown in Table A.6.

W1: residue({1, -2, -3, 4}, {1, 4, -5})

W1: Residue({1, -2, -3, 4}, {1, 4, -5})			
	1: Residues	2: Poles	3: Direct Terms
1:	26.000000	-5.000000	1.000000
2:	0.000000	1.000000	-6.000000
3:			
4:			

Table A.6: Showing the results for Eq: A17

Eq:A17 therefore becomes,

$$\frac{P(x)}{Q(x)} = \frac{x^3-2x^2-3x+4}{x^2+4x-5} = (x-6) + \frac{26}{(x+5)(x-1)} \qquad \dots A18$$

We can also consider an example in the Z domain when the order of the nominator is greater than the denominator.

Example 7: Consider the expression,

$$H(z) = \frac{1-0.23z^{-1}+0.056z^{-2}-0.2014z^{-3}+0.042z^{-4}}{1-0.074z^{-2}-0.83z^{-3}} \qquad \dots A19$$

In this expression the nominator is of order 4 whereas the denominator is of order 3. Using the *residuez* command,

---

W1: residuez({1, -.23, .056, -.2014, .042}, {1, -.074, -.83})

Table A.7 is the result of this command.

W1: residuez({1, -.23, .056, -.2014, .042}, {1, -.074, -.83})

	1: Residues	2: Poles	3: Direct Terms
1:	0.819661	-0.874794	-0.150473
2:	0.330812	0.948794	0.247162
3:			-0.050602
4:			

Table A.7: The factorised results of Eq:A19

From the information in Table A.7, Eq:A19 becomes,

$$H(z) = (z - 0.050602) - \frac{0.150473(z - 0.819661)}{(z + 0.874794)}$$
$$+ \frac{0.247162(z - 0.330812)}{(z - 0.948794)} \qquad \dots A20$$

You will observe from Eq:A20 there are two poles which are both real and to observe the locations of the zeros and poles of Eq:A19 the DADiSP *zplane* command can be used as illustrated in Example 5.

Index

224

225

226

227